大数据科学丛书

Spark SQL 大数据实例开发教程

王家林　段智华　编著

机 械 工 业 出 版 社

Spark SQL是Spark生态环境中最核心和最基础的组件，是掌握Spark的关键所在。本书完全从企业级开发的角度出发，结合多个企业级应用案例，深入剖析Spark SQL。全书共分为8章，包括：认识Spark SQL、DataFrame原理与常用操作、Spark SQL操作多种数据源、Parquet列式存储、Spark SQL内置函数与窗口函数、Spark SQL UDF与UDAF、Thrift Server、Spark SQL综合应用案例。

本书可以使读者对Spark SQL有深入彻底的理解，本书适合于Spark学习爱好者，是学习Spark SQL的入门和提高教材，也是Spark开发工程师开发过程中查阅Spark SQL的案头手册。

图书在版编目（CIP）数据

Spark SQL大数据实例开发教程/王家林等编著.—北京：机械工业出版社，2017.9

（大数据科学丛书）

ISBN 978-7-111-59197-9

Ⅰ.①S… Ⅱ.①王… Ⅲ.①数据处理软件－教材 Ⅳ.①TP274

中国版本图书馆CIP数据核字（2018）第033108号

机械工业出版社（北京市百万庄大街22号 邮政编码100037）
策划编辑：王 斌　责任编辑：王 斌
责任校对：张艳霞　责任印制：孙 炜
北京中兴印刷有限公司印刷

2018年3月第1版·第1次印刷
184mm×260mm·16.5印张·398千字
0001-3000册
标准书号：ISBN 978-7-111-59197-9
定价：59.00元

凡购本书，如有缺页、倒页、脱页，由本社发行部调换

电话服务　　　　　　　　　　　网络服务
服务咨询热线：010-88361066　　机工官网：www.cmpbook.com
读者购书热线：010-68326294　　机工官博：weibo.com/cmp1952
　　　　　　　010-88379203　　金 书 网：www.golden-book.com
封面无防伪标均为盗版　　　　　　教育服务网：www.cmpedu.com

前　言

"Use of MapReduce engine for Big Data projects will decline, replaced by Apache Spark."
MapReduce 计算模型的使用会越来越少，最终将被 Apache Spark 所取代。

——Hadoop 之父 Doug Cutting

写作背景

Spark 是一个快速大规模数据处理的通用引擎。它给 Java、Scala、Python 和 R 等语言提供了高级 API，并基于统一抽象的 RDD（弹性分布式数据集），逐渐形成了一套自己的生态系统。这个生态系统主要包括负责 SQL 和结构化数据处理的 Spark SQL、负责实时流处理的 Spark Streaming、负责图计算的 Spark GraphX 以及机器学习子框架 Mlib。Spark 在处理各种场景时，提供给用户统一的编程体验，可极大地提高编程效率。

Hive 是运行在 Hadoop 上的 SQL on Hadoop 工具，它的推出是为了给熟悉 RDBMS 但又不理解 MapReduce 的技术人员提供快速上手的工具，但是 MapReduce 在计算过程中消耗大量 I/O 资源，降低了运行效率。为了提高 SQL on Hadoop 的效率，Shark 出现了，它使得 SQL on Hadoop 的性能比 Hive 有了 10～100 倍的提高。但 Shark 对于 Hive 的过度依赖（如采用 Hive 的语法解析器、查询优化器等），制约了 Spark 的发展，所以提出了 Spark SQL 项目，Spark SQL 抛弃 Shark 原有的弊端，又汲取了 Shark 的一些优点，如内存列存储（In-Memory Columnar Storage）、Hive 的兼容性等，由于摆脱了对 Hive 的依赖性，Spark SQL 在数据兼容、性能优化、组件扩展等方面的性能都得到了极大的提升。

Spark SQL 是 Spark 生态环境中最核心和最基础的组件，是掌握 Spark 的关键所在。由于目前市场上介绍 Spark 技术的书籍比较少，尤其是单独讲解 Spark SQL 的书更是凤毛麟角，我们特意编写了这本理论和实战相结合的 Spark SQL 书籍，在介绍 Spark SQL 核心技术的同时又配备了丰富的示例，同时还穿插了源代码的分析，使读者能从更深层次来把握 Spark SQL 的核心技术。

内容速览

本书完全从企业级开发的角度出发，结合多个企业级应用案例，深入剖析 Spark SQL。

全书一共分为 8 章，主要内容概括如下：

第 1 章认识 Spark SQL，引领读者了解 Spark SQL 的基础知识，接下来的第 2 章至第 7 章，结合实战案例，引导读者掌握 Spark SQL 的核心知识，这 6 章内容分别为：DataFrame 原理与常用操作、Spark SQL 操作多种数据源、Parquet 列式存储、Spark SQL 内置函数与窗口函数、Spark SQL UDF 与 UDAF、Thrift Server；本书的最后部分，第 8 章 Spark SQL 综合应用案例归纳并综合运用了全部 Spark SQL 知识点，是深入理解 Spark SQL 的经典案例。

本书可以使读者对 Spark SQL 有深入的理解，是 Spark 爱好者用来学习 Spark SQL 的理想教程，也是 Spark 开发工程师在开发过程中可随时查阅的案头手册。

本书作者

本书由王家林和段智华编写。

预备知识

在学习本书之前读者需要熟悉基本的 Linux 命令及 Java、Scala 语言，掌握基本的 Spark 知识架构，能够搭建 Spark 集群环境。

致谢

在本书编写的过程中，作者参考了很多网络上的书籍和博客，在此谢谢各位作者，正是你们的无私奉献，才推动了 Spark 技术的快速发展。

特别感谢"小小"同学为本书的编写提供的各种协调和热心帮助。

由于笔者能力有限，书中难免存在错误或表达不准确的内容，恳请大家批评指正，希望大家一起努力使 Spark 技术在大数据计算领域里推广开来。

作者

2017.9

目　　录

前言

第1章　认识 Spark SQL … 1
1.1　Spark SQL 概述 … 1
1.1.1　Spark SQL 与 DataFrame … 1
1.1.2　DataFrame 与 RDD 的差异 … 2
1.1.3　Spark SQL 的发展历程 … 3
1.2　从零起步掌握 Hive … 4
1.2.1　Hive 的本质是什么 … 4
1.2.2　Hive 安装和配置 … 5
1.2.3　使用 Hive 分析搜索数据 … 12
1.3　Spark SQL on Hive 安装与配置 … 15
1.3.1　安装 Spark SQL … 15
1.3.2　安装 MySQL … 18
1.3.3　启动 Hive Metastore … 21
1.4　Spark SQL 初试 … 21
1.4.1　通过 spark–shell 来使用 Spark SQL … 21
1.4.2　Spark SQL 的命令终端 … 24
1.4.3　Spark 的 Web UI … 25
1.5　本章小结 … 26

第2章　DataFrame 原理与常用操作 … 27
2.1　DataFrame 编程模型 … 27
2.2　DataFrame 基本操作实战 … 28
2.2.1　数据准备 … 28
2.2.2　启动交互式界面 … 30
2.2.3　数据处理与分析 … 31
2.3　通过 RDD 来构建 DataFrame … 44
2.4　缓存表（列式存储） … 47
2.5　DataFrame API 应用示例 … 48
2.6　本章小结 … 79

第3章　Spark SQL 操作多种数据源 … 80
3.1　通用的加载/保存功能 … 80
3.1.1　Spark SQL 加载数据 … 80
3.1.2　Spark SQL 保存数据 … 82
3.1.3　综合案例——电商热销商品排名 … 82

3.2　Spark SQL 操作 Hive 示例 …… 87
3.3　Spark SQL 操作 JSON 数据集示例 …… 91
3.4　Spark SQL 操作 HBase 示例 …… 92
3.5　Spark SQL 操作 MySQL 示例 …… 97
　　3.5.1　安装并启动 MySQL …… 97
　　3.5.2　准备数据表 …… 98
　　3.5.3　操作 MySQL 表 …… 101
3.6　Spark SQL 操作 MongoDB 示例 …… 111
　　3.6.1　安装配置 MongoDB …… 111
　　3.6.2　启动 MongoDB …… 113
　　3.6.3　准备数据 …… 114
　　3.6.4　Spark SQL 操作 MongoDB …… 116
3.7　本章小结 …… 122

第 4 章　Parquet 列式存储　123
4.1　Parquet 概述 …… 123
　　4.1.1　Parquet 的基本概念 …… 123
　　4.1.2　Parquet 数据列式存储格式应用举例 …… 125
4.2　Parquet 的 Block 配置及数据分片 …… 128
　　4.2.1　Parquet 的 Block 的配置 …… 129
　　4.2.2　Parquet 内部的数据分片 …… 129
4.3　Parquet 序列化 …… 129
　　4.3.1　Spark 实施序列化的目的 …… 130
　　4.3.2　Parquet 两种序列化方式 …… 130
4.4　本章小结 …… 131

第 5 章　Spark SQL 内置函数与窗口函数　132
5.1　Spark SQL 内置函数 …… 132
　　5.1.1　Spark SQL 内置函数概述 …… 132
　　5.1.2　Spark SQL 内置函数应用实例 …… 133
5.2　Spark SQL 窗口函数 …… 143
　　5.2.1　Spark SQL 窗口函数概述 …… 143
　　5.2.2　Spark SQL 窗口函数分数查询统计案例 …… 145
　　5.2.3　Spark SQL 窗口函数 NBA 常规赛数据统计案例 …… 154
5.3　本章小结 …… 161

第 6 章　Spark SQL UDF 与 UDAF　162
6.1　UDF 概述 …… 162
6.2　UDF 示例 …… 162
　　6.2.1　Hobby_count 函数 …… 163
　　6.2.2　Combine 函数 …… 164
　　6.2.3　Str2Int 函数 …… 165

目录

- 6.2.4 Wsternstate 函数 ········ 167
- 6.2.5 ManyCustomers 函数 ········ 168
- 6.2.6 StateRegion 函数 ········ 169
- 6.2.7 DiscountRatio 函数 ········ 170
- 6.2.8 MakeStruct 函数 ········ 171
- 6.2.9 MyDateFilter 函数 ········ 172
- 6.2.10 MakeDT 函数 ········ 174
- 6.3 UDAF 概述 ········ 176
- 6.4 UDAF 示例 ········ 176
  - 6.4.1 ScalaAggregateFunction 函数 ········ 176
  - 6.4.2 GeometricMean 函数 ········ 180
  - 6.4.3 CustomMean 函数 ········ 183
  - 6.4.4 BelowThreshold 函数 ········ 186
  - 6.4.5 YearCompare 函数 ········ 188
  - 6.4.6 WordCount 函数 ········ 194
- 6.5 本章小结 ········ 198

第7章 Thrift Server ········ 199

- 7.1 Thrift 概述 ········ 199
  - 7.1.1 Thrift 的基本概念 ········ 199
  - 7.1.2 Thrift 的工作机制 ········ 201
  - 7.1.3 Thrift 的运行机制 ········ 201
  - 7.1.4 一个简单的 Thrift 实例 ········ 203
- 7.2 Thrift Server 的启动过程 ········ 206
  - 7.2.1 Thrift Sever 启动详解 ········ 207
  - 7.2.2 HiveThriftServer2 类的解析 ········ 212
- 7.3 Beeline 操作 ········ 215
  - 7.3.1 Beeline 连接方式 ········ 215
  - 7.3.2 在 Beeline 中进行 SQL 查询操作 ········ 218
  - 7.3.3 通过 Web 控制台查看用户进行的操作 ········ 220
- 7.4 Thrift Server 应用示例 ········ 221
  - 7.4.1 示例源代码 ········ 221
  - 7.4.2 关键代码行解析 ········ 222
  - 7.4.3 测试运行 ········ 224
  - 7.4.4 运行结果解析 ········ 227
  - 7.4.5 Spark Web 控制台查看运行日志 ········ 227
- 7.5 本章小结 ········ 228

第8章 Spark SQL 综合应用案例 ········ 229

- 8.1 综合案例实战——电商网站日志多维度数据分析 ········ 229
  - 8.1.1 数据准备 ········ 230

8.1.2 数据说明 …………………………………………………………… 230
8.1.3 数据创建 …………………………………………………………… 230
8.1.4 数据导入 …………………………………………………………… 235
8.1.5 数据测试和处理 …………………………………………………… 240
8.2 综合案例实战——电商网站搜索排名统计 …………………………………… 245
8.2.1 案例概述 …………………………………………………………… 245
8.2.2 数据准备 …………………………………………………………… 245
8.2.3 实现用户每天搜索前 3 名的商品排名统计 ……………………… 249
8.3 本章小结 …………………………………………………………………………… 254

第 1 章 认识 Spark SQL

1.1 Spark SQL 概述

Spark SQL 是 Spark 的计算模块之一,它和 Spark 的基础模块 RDD 不一样,是专门用于处理结构化数据的。Spark SQL 为 Spark 提供了更加便利的处理和计算结构化数据的能力。具备以下知识基础,可以让我们更好地了解和使用 Spark SQL,包括:SQL、Spark SQL DataFrame、Spark SQL Dataset(SparkSQL Dataset 在 Spark 2.0 之后才被正式推出,在 Spark 1.6.3 版本的生产环境中还没有大规模应用)。

Spark SQL 既可以使用标准的 SQL 语法,也可以使用 HiveQL 来执行 SQL 的查询和读写,Spark SQL 还可以从已经存在的 Hive 数据仓库中读取数据(关于这方面的配置,我们将在 1.3 节中介绍)。当我们在 Spark 程序中使用 Spark SQL 时,其结果将返回 DataFrame。当然,我们还可以通过交互式命令行或者 JDBC/ODBC 来调用 Spark SQL 的 API。

1.1.1 Spark SQL 与 DataFrame

Spark SQL 是在 Spark 生态系统中除了 Spark Core 之外最大且最受关注组件,如图 1-1 所示。

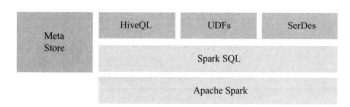

图 1-1 Spark 组件

Spark SQL 具备以下特征:

1)处理一切存储介质和各种格式的数据,同时可以方便地扩展 Spark SQL 的功能来支持更多类型的数据,例如 Kudu。

2)Spark SQL 把数据仓库的计算能力推向了新的高度,不仅包括高效的计算速度(Spark SQL 比 Shark 快了至少一个数量级,而 Shark 比 Hive 快了至少一个数量级,尤其是在 Tungsten Project 成熟以后 Spark SQL 会更加无可匹敌),更为重要的是 Spark SQL 大大提升了数据仓库的计算复杂度(Spark SQL 推出的 DataFrame 可以让数据仓库直接使用机器学习、图计算等复杂的算法库来对数据仓库进行复杂深度数据价值的挖掘)。

3) Spark SQL (DataFrame、DataSet) 不仅是数据仓库的引擎，而且也是数据挖掘的引擎，更为重要的是 Spark SQL 是数据科学计算和分析引擎。

4) Hive + Spark SQL + DataFrame 组成了目前在国内的大数据主流技术组合：
- Hive：负责低成本的数据仓库存储。
- Spark SQL：负责高速的计算。
- DataFrame：负责复杂的数据挖掘。

1.1.2　DataFrame 与 RDD 的差异

DataFrame 是一个分布式的面向列组成的数据集，在概念上类似于一张关系型数据库中的表，但是在 Spark 计算引擎下，有了更高效的优化。DataFrame 的组成来源很广泛，例如：结构化文件、Hive 中的 Table、外部数据库，或者由 RDD 转换而来。DataFrame 的 API 可以在 Scala、Java、Python、R 中使用（Spark 支持以上 4 种语言的开发）。

在 R 和 Python 中都有 DataFrame，但 Spark SQL 中的 DataFrame 从形式上看最大的不同点在于，其天生是分布式的。从学习的角度讲，我们可以简单地把 Spark SQL 中的 DataFrame 抽象成是一个分布式的表，其形式如表 1-1 所示。

表 1-1　DataFrame 被抽象成分布式的表

姓名	年龄	电话
String	Int	Long
String	Int	Long
String	Int	Long
…	…	…
String	Int	Long
String	Int	Long

而在 RDD 中，DataFrame 的形式如表 1-2 所示。

表 1-2　RDD 中 Dataframe 的表现形式

Record
Record
Record
Record
Record
Record
Record

数据的表现在 RDD 和 DataFrame 之间有两点根本的差异：

1) RDD 是以 Record 为单位的，SparkSQL 在优化的时候无法了解 Record 内部的细节，所以也就无法进行更深度的优化，这极大地限制了 Spark SQL 性能的提升。

2）DataFrame 是以列为单位的，包含每个 Record 的元数据信息，也就是说 DataFrame 在优化时基于列内部的优化，而不是像 RDD 一样，只能够基于行来进行优化。

1.1.3　Spark SQL 的发展历程

SparkSQL 发展历程图如图 1-2 所示。

图 1-2　Spark SQL 发展历程图

SparkSQL 是从 Shark 发展而来的，2014 年 7 月 1 日的 Spark Summit 上，Databricks 宣布终止对 Shark 的开发，转向到 Spark SQL 上。Databricks 表示，Spark SQL 将涵盖 Shark 的所有特性，用户可以从 Shark 0.9 进行无缝的升级。同时 Databricks 推出 Spark SQL 和 Hiveon Spark。其中 Spark SQL 是为 Spark 设计的一代新的 SQL 引擎，作为 Spark 生态中的一员继续发展，而不再受限于 Hive，只是兼容 Hive；而 Hive on Spark 是将 Spark 作为一个替代执行引擎提供给 Hive 的，从而为已经存在的 Hive 用户提供了一个迁往 Spark 的途径。

下面简单地介绍一下 Shark 及 Shark 项目终止的原因。

Shark 发布时，Hive 可以说是 SQL on Hadoop 的唯一选择，负责将 SQL 编译成可扩展的 MapReduce 作业。鉴于 Hive 的性能及与 Spark 的兼容性，Shark 项目由此而生。

Shark 通过 Hive 的 HQL（Hibernate Query Language）解析，把 HQL 翻译成 Spark 上的 RDD 操作，然后通过 Hive 的元数据获取数据库里的表信息，实际 HDFS 上的数据和文件，会由 Shark 获取并放到 Spark 上运算。

Shark 的最大特性就是速度快以及与 Hive 的完全兼容，且可以在 Shell 模式下使用 API，把 HQL 得到的结果集继续在 Scala 环境下运算，支持自己编写简单的机器学习或简单分析处理函数，对 HQL 结果进一步分析计算。

但是 Shark 更多的是对 Hive 的改造，替换了 Hive 的物理执行引擎，因此会有一个很快的速度。然而，不容忽视的是，Shark 继承了大量的 Hive 代码，因此给进一步对其优化和维护带来了大量的麻烦。随着性能优化和先进分析整合的进一步加深，基于 MapReduce 设计的部分无疑成为整个项目的瓶颈。因此，为了更好地发展，给用户提供一个更好的体验，Databricks 宣布终止 Shark 项目，从而将更多的精力放到 Spark SQL 上。

Section 1.2 从零起步掌握 Hive

Hive 是基于 Hadoop 构建的一套数据仓库分析系统，提供了丰富的 SQL 查询方式来分析存储在 Hadoop 分布式文件系统中的数据，可以将结构化的数据文件映射为一张数据库表，并提供完整的 SQL 查询功能，可以将 SQL 语句转换为 MapReduce 任务进行运行，通过自己的 SQL 去查询分析需要的内容，这套 SQL 简称 HiveQL（Hive SQL）。

1.2.1 Hive 的本质是什么

Hive 是分布式数据仓库，同时又是查询引擎，所以 Spark SQL 取代的只是 Hive 查询引擎，在企业实际生产环境下，Hive + Spark SQL 是目前最为经典的数据分析组合，Hive 本身就是一个简单单机版本的软件，主要负责：

1) 把 HQL 翻译成 Mapper - Reducer - Mapper 的代码，并且可能产生很多 MapReduce 的 Job。

2) 把生产的 MapReduce 代码及资源打成 JAR 包，并自动发布到 Hadoop 集群中运行。

Hive 本身的架构如图 1-3 所示。

Hive 架构主要有下面 4 个部分组成：

1) 驱动程序（Driver）：负责编译、优化、执行。

Hive 的入口是 Driver，执行的 SQL 语句首先提交到 Driver，然后调用编译器（Compiler）解释驱动，最终解释成 MapReduce 任务执行，最后将结果返回。Driver 调用编译器处理 HiveQL（Hive SQL），可能是一条 DDL、DML 或查询语句。编译器将字符串转化为策略（Plan）。策略仅由元数据操作和 HDFS 操作组成，元数据操作只包含 DDL 语句，HDFS 操作只包含 LOAD 语句。具体流程为：解析→语义分析→逻辑策略生成→优化→执行。

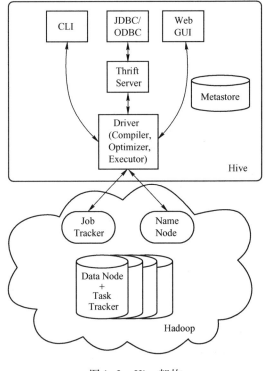

图 1-3 Hive 架构

2) 3 种服务模式：Hive 驱动对外提供了 3 种服务模式，分别是：

- CLI：即 Hive 命令行模式，通过命令行终端来直接操作 Hive。
- Web GUI：即 Hive 的 Web 模式，通过浏览器来访问 Hive。
- Hive 的远程服务模式：通过 Thrift Server 的支持，远程 Client 程序可以通过 JDBC/ODBC 方式来访问连接 Hive，这也是程序员最需要的方式。

3）Metastore：元数据存储。

Hive 将元数据存储在 RDBMS 中，一般常用 MySQL 和 Derby。默认情况下，Hive 元数据保存在内嵌的 Derby 数据库中，只能允许一个会话连接，只适合简单的测试。实际生产环境中不适用，为了支持多用户会话，则需要一个独立的元数据库，使用 MySQL 作为元数据库，Hive 内部对 MySQL 提供了很好的支持。

4）Hadoop：Hadoop 是 Hive 的运行基石。

Hive 安装依赖 Hadoop 的集群，Hive 运行在 Hadoop 上，用 HDFS 进行存储，利用 MapReduce 进行计算。

1.2.2 Hive 安装和配置

要安装 Hive，首先要去 Hive 的官网（hive.apache.org）下载安装包（本书选择 Hive-1.2.1 版本）。在这里要说明一下，官网上的安装包不带基于 Web 的图形化的查询工具，如果需要的话，可以自行下载源码包，自己编译打包，生成基于 Web 的图形化查询工具，然后进行部署。这个时候就可以使用 Web 的管理工具来查询数据仓库中的数据了。

我们知道 Hive 是运行在 Hadoop 之上的，所以在安装 Hive 之前，我们要先安装好 Hadoop 环境，Hadoop 可以是单机环境，也可以是伪分布环境，还可以是集群环境，我们采用的是 Hadoop-2.6.0 版的集群环境。

下面介绍 Hive 的安装模式，Hive 有 3 种安装模式，分别是嵌入模式、本地模式、远程模式。

（1）嵌入模式安装

在这种模式下，Hive 的元数据信息被存储在 Hive 自带的 Derby 数据库中。Hive 的嵌入模式有很大的局限，在同一时间，Hive 只允许创建一个连接，这意味着，这个时候只能有一个人可以操作 Hive，这种模式一般只适用于做演示使用。

（2）本地模式安装

实际上本地模式和嵌入模式很相似，这个时候 Hive 的元数据存储在本地另外的数据库当中，通常我们使用 MySQL 作为 Hive 的元数据数据库。在这种模式下，允许多个用户同时连接，这种模式一般用在我们的开发和测试中。

（3）远程模式安装

一般生产环境采用的都是远程模式，在远程模式下，Hive 和元数据数据库 MySQL（一般是 MySQL，本书中同样采用的是 MySQL，如果不做特殊的说明，本章所有的元数据数据库都指的是 MySQL 数据库）运行在不同的机器上，且操作系统也可能不一样。在这种模式下，Hive 允许多用户同时连接。

下面分别介绍 Hive 的 3 种模式下的安装步骤。

安装环境说明：

- 操作系统：Ubuntu14.04 LTS。
- 软件版本：apache-hive-1.2.1-bin.tar.gz。
- Hadoop：hadoop-2.6.0 集群模式。

1. Hive 的嵌入模式安装步骤

1）解压安装包，并且将安装包复制到指定的目录，假设为：/opt/software。

```
tar -zxvf apache-hive-1.2.1-bin.tar.gz -C/opt/software
```

解压后进入到该目录中，可以看到下列子目录：

```
bin        examples   lib        NOTICE          RELEASE_NOTES.txt
conf       hcatalog   LICENSE    README.txt      scripts
```

简单介绍一下 Hive 的目录结构。
- bin 目录：存放的是一些可执行文件，比如 Hive 常用的一些指令等。
- conf 目录：存放的是 Hive 的配置文件，比如 Hive 的元数据存储信息等的配置，都在这个文件目录中。
- examples 目录：存放 Hive 官方提供的一些案例程序。
- lib 目录：存放 Hive 的一些 JAR 包，通过这些 JAR 包，我们就可以调用 Hive 的指令来执行操作了。

2）将 Hive 安装目录下的 lib 目录中的 jline-2.12.jar 文件复制到 Hadoop 安装目录下的 share/hadoop/yarn/lib 目录中，否则在启动 Hive 的时候会报错。

```
cp jline-2.12.jar /opt/software/hadoop-2.6.0/share/hadoop/yarn/lib
```

3）到 Hive 安装目录下的 bin 目录中，执行 ./hive 命令启动 Hive，在启动 Hive 的同时，Hive 会自动创建一个 Derby 数据库作为元数据存储介质，至此，Hive 的嵌入模式安装完成。如果见到下面的提示，说明 Hive 已经安装成功了。

```
hadoop@hadoop:/opt/software/hive/bin $ ./hive
Logging initialized using configuration in jar:file:/opt/software/hive/lib/hive-common-1.2.1.jar!/
    hive-log4j.properties
hive >
```

接下来可以试试 Hive 是否能够正常使用。如下所示，使用 SHOW DATABASES 查询数据库。

```
hive > SHOW DATABASES;
OK
Default
Time taken:1.58 seconds,Fetched:1 row(s)
hive >
```

现在也可以输入其他的命令，就像操作数据库一样，可以创建表、插入数据、删除数据等。

2. Hive 的本地模式和远程模式安装

由于 Hive 的本地模式和远程模式非常相似，所以这里就介绍 Hive 的远程模式安装。远程模式安装意味着 Hive 的元数据存储在远程机器上，远程机器可以是 Linux 系统，也可以是 Windows 系统。这个根据实际生产环境所决定。

在本章中，我们使用 MySQL 作为 Hive 的元数据数据库，远程机器操作系统是 Ubuntu14.04LTS 版。关于 MySQL 的安装，可以参考本章的第 1.3.2 小节。

下面将分为两个阶段进行安装工作。

第一阶段：进行与 MySQL 相关的准备工作。

1）在进行 Hive 的远程模式安装之前，先登录 MySQL 查看已有的数据库的信息。

```
hadoop@ hadoop:/opt/software $ mysql -u root -p
Enter password:
Welcome to the MySQL monitor.    Commands end with;or \g.

mysql > SHOW DATABASES;
+--------------------+
| Database           |
+--------------------+
| information_schema |
| hive               |
| mysql              |
| performance_schema |
+--------------------+
4 rows in set (0.00 sec)
```

结果显示，登录 MySQL 数据库查询到已有四个数据库，分别为：information_schema、hive、mysql、performance_schema。

2）创建元数据信息库。

在上面列表中，如果发现名称为 Hive 的数据库不存在，那么可以手动新建一个名为 hive 的数据库，用来存储 Hive 数据仓库中的元数据信息（当然也可以不手动创建它，因为 Hive 的相应配置可以支持自动创建元数据信息库）。

下面先手工新建一个名称为 hive 的数据库：

```
mysql > CREATE DATABASE hive;
Query OK,1 row affected (0.00 sec)
```

这个时候我们已经成功地创建了数据库，目前里面暂时是空的。

```
mysql > USE hive;
Database changed
mysql > SHOW TABLES;
Empty set (0.00 sec)
```

3）MySQL 驱动程序准备。

因为我们使用 MySQL 作为元数据数据库，所以还需要把 MySQL 的驱动放到 Hive 安装目录下的 lib 子目录中。MySQL 驱动程序可以去官网上下载，我们选择的是 mysql-connector-java-5.1.39-bin.jar 这个文件包。检查该包是否存在于 lib 目录中：

```
root@ master:/opt/software/apache-hive-1.2.1-bin/lib# ll my*
-rw-r--r-- 1 root root 855948 Nov  8 15:23 mysql-connector-java-5.1.39-bin.jar
```

如上所示，查询到 mysql-connector-java-5.1.39-bin.jar 已位于 lib 目录之下。

4) MySQL 的访问账号配置。

因为我们是使用 root 账号来做演示的，所以需要让 root 账号可以被远程连接，同时删除所有的匿名用户。下面是具体的操作步骤：

```
hadoop@ hadoop:~ $ /usr/bin/mysql_secure_installation

NOTE:RUNNING ALL PARTS OF THIS SCRIPT IS RECOMMENDED FOR ALL MySQL
     SERVERS IN PRODUCTION USE!    PLEASE READ EACH STEP CAREFULLY!

In order to log into MySQL to secure it,we'll need the current
password for the root user.   If you've just installed MySQL,and
you haven't set the root password yet,the password will be blank,
so you should just press enter here.

Enter current password for root (enter for none):
OK,successfully used password,moving on...

Setting the root password ensures that nobody can log into the MySQL
root user without the properauthorisation.

You already have a root password set,so you can safely answer 'n'.

Change the root password? [Y/n] n
```

当提示是否修改 root 密码时，由于已经设置好了，所以这里选择 n。接下来会显示下面这个提示界面：

```
By default,a MySQL installation has an anonymous user,allowing anyone
to log into MySQL without having to have a user account created for
them. This is intended only for testing,and to make the installation
go a bit smoother.   You should remove them before moving into a
production environment.

Remove anonymous users? [Y/n] y
```

询问是否删除匿名用户，这里选择 Y 确认，之后的每一次询问都选择 Y。

```
By default,MySQL comes with a database named 'test'that anyone can
access.   This is also intended only for testing,and should be removed
before moving into a production environment.
Remove test database and access to it? [Y/n] y
 - Dropping test database...
```

第1章 认识Spark SQL

```
ERROR 1008 (HY000) at line 1:Can't drop database 'test';database doesn't exist
 ... Failed!    Not critical,keep moving...
 - Removing privileges on test database...
 ... Success!

Reloading the privilege tables will ensure that all changes made so far
will take effect immediately.

Reload privilege tables now? [Y/n] y
 ... Success!

Cleaning up...

All done!  If you've completed all of the above steps,your MySQL
installation should now be secure.

Thanks for using MySQL!
```

5)给root用户授权,使root用户可以被远程连接。

```
mysql > GRANT ALL PRIVILEGES ON *.* TO 'root'@'%' IDENTIFIED BY 'root' WITH GRANT
     OPTION;
Query OK,0 rows affected (0.00 sec)
```

授权成功后更新权限。

```
mysql > FLUSH PRIVILEGES;
Query OK,0 rows affected (0.00 sec)
```

6)查看user表中的所有用户信息,结果如下所示,查询出了root、debian-sys-maint用户以及相应的主机名、地址。

```
mysql > select user,host from mysql.user;
+-------------------+-----------+
| user              | host      |
+-------------------+-----------+
| root              | %         |
| root              | 127.0.0.1 |
| root              | ::1       |
| debian-sys-maint  | localhost |
| root              | localhost |
+-------------------+-----------+
5 rows in set (0.00 sec)
```

第二阶段:MySQL准备工作完毕,进入Hive远程模式安装环节。
1)首先,解压Hive的安装包到指定目录下。

```
tar -zxvf apache-hive-1.2.1-bin.tar.gz    -C /opt/software
```

这样就将 Hive 的安装目录解压并复制到了 /opt/software 目录下了。然后，进入 Hive 安装目录下的 conf 目录。

2）查看 Hive 的 conf 目录。

进入 conf 目录后，我们可以看到 6 个默认的配置信息文件，是 Hive 配置文件的模板。分别为：日志配置文件（beeline-log4j.properties.template、hive-log4j.properties.template、hive-exec-log4j.properties.template）、Hive 缺省配置文件（hive-default.xml.template）、Hive 环境配置文件（hive-env.sh.template），以及项目依赖项配置文件（ivysettings.xml）。

```
cd /opt/software/apache-hive-1.2.1-bin/conf
beeline-log4j.properties.template      hive-log4j.properties.template
hive-default.xml.template              ivysettings.xml
hive-env.sh.template
```

3）配置 hive-site.xml。

在 Hive 的 conf 目录下创建 hive-site.xml（可以参考 hive-default.xml.template 文件创建，也可以自己手动新建一个 hive-site.xml 文件），命令如下。

```
vi hive-site.xml
```

然后在该文件中输入以下内容：

```
<?xml version="1.0" encoding="UTF-8" standalone="no"?>
<?xml-stylesheet type="text/xsl" href="configuration.xsl"?>
<configuration>
<property>
<name>javax.jdo.option.ConnectionURL</name>       <!--指定 mysql 数据库地址-->
<value>jdbc:mysql://192.168.0.38:3306/hive</value>
</property>
<property>
<name>javax.jdo.option.ConnectionDriverName</name>   <!--指定数据库驱动器名称-->
<value>com.mysql.jdbc.Driver</value>
</property>
<property>
<name>javax.jdo.option.ConnectionUserName</name>    <!--指定数据库用户名-->
<value>root</value>
</property>
<property>
<name>javax.jdo.option.ConnectionPassword</name>    <!--指定数据库用户密码-->
<value>root</value>
</property>
</configuration>
```

这样，就完成了 hive-site.xml 的配置。
4）验证 Hive 是否安装成功。
进入 Hive 的 bin 目录，启动 Hive，命令如下：

```
./hive
```

执行完上面的指令后，如果在 SHELL 中看到如下提示，那么就表示已经安装成功了：

```
hadoop@ hadoop:/opt/software/apache-hive-1.2.1-bin/bin $ ./hive

Logging initialized using configuration in jar:file:/opt/software/apache-hive-1.2.1-bin/lib/hive-
    common-1.2.1.jar!/hive-log4j.properties
hive >
```

这个时候可以去 MySQL 中查看数据库列表，发现存在了一个名为 hive 的数据库：

```
mysql > show databases;
+--------------------+
| Database           |
+--------------------+
| information_schema |
| hive               |
| mysql              |
| performance_schema |
+--------------------+
4 rows in set (0.00 sec)
```

然后可以查看该数据库中有哪些表。

```
mysql > use hive

mysql > show tables;
+---------------------------+
| Tables_in_hive            |
+---------------------------+
| BUCKETING_COLS            |
| CDS                       |
| COLUMNS_V2                |
| DATABASE_PARAMS           |
| DBS                       |
| FUNCS                     |
| FUNC_RU                   |
| GLOBAL_PRIVS              |
```

```
| PARTITIONS                  |
| PARTITION_KEYS              |
| PARTITION_KEY_VALS          |
| PARTITION_PARAMS            |
| PART_COL_STATS              |
| ROLES                       |
| SDS                         |
| SD_PARAMS                   |
| SEQUENCE_TABLE              |
| SERDES                      |
| SERDE_PARAMS                |
| SKEWED_COL_NAMES            |
| SKEWED_COL_VALUE_LOC_MAP    |
| SKEWED_STRING_LIST          |
| SKEWED_STRING_LIST_VALUES   |
| SKEWED_VALUES               |
| SORT_COLS                   |
| TABLE_PARAMS                |
| TAB_COL_STATS               |
| TBLS                        |
| VERSION                     |
+-----------------------------+
```

从上面的信息中可以看到，Hive 已经在 MySQL 中创建了用来存储元数据信息的相关表。

1.2.3 使用 Hive 分析搜索数据

接下来，我们演示如何使用 Hive 处理数据，主要步骤如下：
- 准备数据。
- 了解要处理的数据。
- 将数据导入到 Hive 中。
- 开始使用 Hive 来分析搜索数据。

1. 数据准备

在使用 Hive 分析数据之前，首先需要准备数据，本示例演示使用的数据来自搜狗实验室，可以自行去搜狗实验室官方网站（http://www.sogou.com/labs/）下载：用户查询日志。该数据格式如图 1-4 所示。

2. 认识数据

首先认识一下要分析的数据，数据一共有 6 列，分别是：
- 第一列，搜索时间。
- 第二列，用户 ID。

第1章 认识Spark SQL

```
20111230000005    57375476989eea12893c0c3811607bcf    奇艺高清    1  1  http://www.qiyi.com/
20111230000005    66c5bb7774e31d0a22278249b26bc83a    凡人修仙传   3  1  http://www.booksky.org/BookDetail.aspx?
                                                                       BookID=1050804&Level=1
20111230000007    b97920521c78de70ac38e3713f524b50    本本联盟    1  1  http://www.bblianmeng.com/
```

图 1-4 要分析的数据

- 第三列，用户在搜索框中输入的搜索内容。
- 第四列，搜索内容出现在搜索结果页面中的第几行。
- 第五列，表示用户单击的是搜索出来的页面上的第几行。
- 第六列，表示用户单击的超链接。

3. 数据导入

准备好数据之后，需要将数据导入到 Hive 数据仓库中。

首先启动 Hive，然后在里面创建一个名为 hive 的数据库，使用该数据库，并在这个数据库中创建一个表，命名为 sogouQ1，具体操作如下：

1）创建一个名为 hive 的数据库。

```
hive > create database hive;
OK
Time taken:0.395 seconds
```

2）使用该数据库。

```
hive > use hive;
OK
```

3）创建一张名为 SogouQ1 的表。

```
hive > create tableSogouQ1(ID string,websesion string,word string,s_seq int,c_seq int,website string)
row format delimited fields terminated by '\t' lines terminated by '\n';
OK
Time taken:0.509 seconds
```

4）将本地数据导入到 SogouQ1 表中。

```
hive > load data local inpath '/opt/data/SogouQ1.txt' into table SogouQ1;
Loading data to table hive.sogouq1
Table hive.sogouq1 stats:[numFiles=1,totalSize=108750574]
OK
Time taken:2.602 seconds
```

现在数据已经准备完毕，需要注意的是，数据的存放包括两种方式：数据存放在本地的磁盘文件中；数据存放在 Hadoop 集群中的 HDFS 分布式文件系统中。如果数据在 HDFS 上，那么我们在导入数据的时候，需要去掉 load 语句中的 local 关键字。另外，Hive 的表分为内部表和外部表，可以简单地理解为，内部表是数据存储在 Hive 数据仓库中的表，而外部表则是数据不存在 Hive 的元数据中。对于内部表来说，在我们删除表的同时，数据也被删除了，而在删除外部表的时候，不会删除元数据。

4. 使用 Hive 分析搜索数据

1）统计 SogouQ1 表的记录数。

该示例统计 SogouQ1 表的记录总数，同时也可以了解 Hive 的具体运行过程。

```
hive > select count( * ) from SogouQ1;
Query ID = hadoop_20160516102827_6e05496e-dab9-4f41-93f4-2d8e30ad015d
Total jobs = 1
Launching Job 1 out of 1
Number of reduce tasks determined at compile time:1
In order to change the average load for a reducer (in bytes):
   set hive.exec.reducers.bytes.per.reducer = < number >
In order to limit the maximum number of reducers:
   set hive.exec.reducers.max = < number >
In order to set a constant number of reducers:
   setmapreduce.job.reduces = < number >
Starting Job = job_1463323240514_0001, Tracking URL = http://hadoop:8088/proxy/application_1463323240514_0001/
Kill Command = /opt/software/hadoop-2.6.0/bin/hadoop job -kill job_1463323240514_0001
Hadoop job information for Stage-1: number of mappers:1; number of reducers:1
2016-05-16 10:28:41,347 Stage-1 map = 0%, reduce = 0%
2016-05-16 10:28:49,197 Stage-1 map = 100%, reduce = 0%, Cumulative CPU 4.68 sec
2016-05-16 10:28:56,695 Stage-1 map = 100%, reduce = 100%, Cumulative CPU 7.74 sec
MapReduce Total cumulative CPU time:7 seconds 740 msec
Ended Job = job_1463323240514_0001
MapReduce Jobs Launched:
Stage-Stage-1: Map:1  Reduce:1  Cumulative CPU:7.74 sec  HDFS Read:108757525 HDFS Write:8 SUCCESS
TotalMapReduce CPU Time Spent:7 seconds 740 msec
OK
1000000
Time taken:31.858 seconds, Fetched:1 row(s)
```

从上面的运行结果中可以看到，SogouQ1 表的数据总共有 100 万条，运行时间 31.858 秒。关于 Hive 的运行过程，这里做一个简单的说明。首先 HQL 语句会被 Hive 转换成 map/reduce 程序，然后通过 hive 自动打包并发布到集群中运行。所以在 Hive 中写 HQL 语句来进行增删改查操作时，其实最终都是通过 map/reduce 程序来完成的。

2）按条件统计 SogouQ1 表中满足条件的记录数。

下面是查询搜索关键字为 baidu 的记录总共有多少条。

```
hive > select count( * ) from SogouQ1 where word like '% baidu%;
Query ID = hadoop_20160517000240_aff7ba60-0d8c-48d3-ae03-d14447ccb7b2
Total jobs = 1
Launching Job 1 out of 1
Number of reduce tasks determined at compile time:1
In order to change the average load for a reducer (in bytes):
```

第1章 认识Spark SQL

```
    set hive. exec. reducers. bytes. per. reducer = <number>
In order to limit the maximum number of reducers:
    set hive. exec. reducers. max = <number>
In order to set a constant number of reducers:
    setmapreduce. job. reduces = <number>
Starting Job = job_1463455574333_0002, Tracking URL = http://master:8088/proxy/application_
    1463455574333_0002/
Kill Command = /opt/software/hadoop-2. 6. 0/bin/hadoop job   - kill job_1463455574333_0002
Hadoop job information for Stage -1:number of mappers:1;number of reducers:1
2016 -05 -17 00:03:05,281 Stage -1 map = 0% ,   reduce = 0%
2016 -05 -17 00:03:16,413 Stage -1 map = 100% ,   reduce = 0% ,Cumulative CPU 6. 03 sec
2016 -05 -17 00:03:23,943 Stage -1 map = 100% ,   reduce = 100% ,Cumulative CPU 7. 97 sec
MapReduce Total cumulative CPU time:7 seconds 970 msec
Ended Job = job_1463455574333_0002
MapReduce Jobs Launched:
Stage - Stage -1: Map:1    Reduce:1    Cumulative CPU:7. 97 sec    HDFS Read:108758478 HDFS
    Write:5 SUCCESS
TotalMapReduce CPU Time Spent:7 seconds 970 msec
OK
4470
Time taken:44. 405 seconds,Fetched:1 row(s)
```

从统计结果可知，包含关键字为 baidu 的记录共有 4470 条，查询时间为 44.405 秒。

3）对 SogouQ1 表进行更复杂的统计。

统计总共有多少条搜索 baidu 且排名和点击率都是第一的记录数。s_seq 表示搜索内容出现在搜索结果页面中的第几行；c_seq 表示用户点击的是搜索出来的页面上的第几行；s_seq = 1 and c_seq = 1 表示搜索内容出现在搜集结果页面的排名是第一行，而且用户点击的就是搜索页面的第一行。

下面是具体的 HQL 语句及执行结果：

```
hive > select count( * ) fromSogouQ1 where s_seq = 1 and c_seq = 1 and word like 'baidu'
result:4124
```

从统计结果可知，符合查询项的记录共有 4124 条。

1.3 Spark SQL on Hive 安装与配置

1.3.1 安装 Spark SQL

要使用 Spark SQL，必须要安装 Spark。Spark 有两种运行模式，一种是 Spark on Yarn，

另一种是 Standalone 模式。本章将带领大家完成 Spark 的 Standalone 模式的安装。

Spark 目前的运行环境只能是 Linux，本书中的运行环境如下：
- Linux 使用的是 Ubuntu14.04 LTS。
- Spark 为 1.6.3 版本。
- Hadoop 是 2.6.0 版本。

简要说一下 Spark 的安装步骤，首先需要一个 Linux 环境，同时已经安装了 Hadoop，然后去 Spark 官网（http://spark.apache.org/）下载 Spark 安装包，将它解压并进行配置。

注意，本章中所有安装配置都是最小化运行配置。

1. Spark 软件包解压

```
tar -zxvf spark-1.6.3-bin-hadoop2.6.gz
```

解压完成后查看 Spark 下的子目录：

```
hadoop@hadoop:/opt/software/spark-1.6.3-bin-hadoop2.6 $ ls
bin conf ec2 lib licenses python README.md sbin
CHANGES.txt data examples LICENSE NOTICE R RELEASE
```

上面的内容就是 Spark 下的子目录，其中 bin 和 sbin 目录中存放 Spark 的基本命令，例如启动集群、启动 spark-sql 等，conf 目录存放 Spark 的配置文件。

2. Spark 的配置

（1）配置 spark-env.sh 配置文件

进入 Spark 的配置目录：

```
cd spark-1.6.3-bin-hadoop2.6/conf
```

输入 vim 命令，编辑 spark-env.sh 文件：

```
vim spark-env.sh
```

在 spark-env.sh 配置文件中加入以下配置项，设置 JAVA、SCALA、HADOOP、HADOOP_CONF 的环境变量：

```
export JAVA_HOME=/opt/software/jdk-1.8.0_65
export SCALA_HOME=/opt/software/scala-2.11.8
export HADOO_HOME=/opt/software/hadoop-2.6.0
export HADOOP_CONF_DIR=$HADOOP_HOME/etc/hadoop
export SPARK_MASTER_IP=master
export SPARK_WORKER_MEMORY=2g
export SPARK_EXCUTOR_MEMORY=2g
export SPARK_DRIVER_MEMORY=2g
export SPARK_WORK_CORES=8
```

SPARK_MASTER_IP 设置 Master 结点地址、SPARK_WORKER_MEMORY 设置 Worker 结

点内存大小、SPARK_EXCUTOR_MEMORY 设置 EXCUTOR 的内存大小、SPARK_DRIVER_MEMORY 设置 DRIVER 的内存大小、SPARK_WORK_CORES 设置 Worker 的内核数。

（2）配置系统环境变量

输入 vim 命令，编辑系统的环境变量文件 ~/.bashrc：

```
Shell:vim ~/.bashrc
```

在环境变量中加入以下配置项：

```
export SPARK_HOME = /opt/software/spark-1.6.3-bin-hadoop2.6
export PATH = $PATH:$SPARK_HOME/bin:$SPARK_HOME/sbin
```

- 设置 Spark 的环境变量 SPARK_HOME。
- 设置 PATH 环境变量，将 SPARK 的 bin，sbin 目录加入 PATH 中。

（3）环境变量生效

在命令行输入 source ~/.bashrc，使修改的 ~/.bashrc 配置文件生效。

```
source ~/.bashrc
```

（4）配置 spark-default.conf 文件

输入 mv 命令，将 spark-default.conf.template 模板更名为 spark-default.conf 文件名。
输入 vim 命令，编辑 spark-default.conf 配置文件。

```
mv spark-default.conf.template spark-default.conf
vim spark-default.conf
```

在 spark-default.conf 配置文件中加入以下配置项：

```
spark.excutor.extraJavaOptions    -XX:+PrintGCDetails -Dkey=value -Dnumbers="one two three"
spark.eventLog.enabled            true
spark.eventLog.dir                hdfs://master:9000/historyserverforSpark
spark.yarn.historyServer.address  master:18080
spark.history.fs.logDirectory     hdfs://master:9000/historyserverforSpark
```

- spark.excutor.extraJavaOptions，配置 Excutor 的 JVM 选项，输出 GC 的详细日志；-Dkey=value 方式指定系统属性；例如 Dnumbers 的值可以设置为"one two three"。
- spark.eventLog.enabled 设置为 true：启动事件日志，记录 Spark 事件日志。
- spark.event-Log.dir，配置事件日志的目录。
- spark.yarn.historyServer.address，设置 historyServer 的地址及端口。
- spark.history.fs.logDirectory，设置历史应用程序的日志目录 URL。

（5）配置 slaves 文件

在 slaves 配置文件中加入 worker 结点的 hostname 主机名：

- 使用 Vim 编辑 slaves 配置文件。
- slaves 文件的每一行配置为 worker 结点的主机名，每行配置一个结点。

```
vim slaves
Worker1
Worker2
Worker3
```

（6）启动 Spark 集群

Spark 已经安装完成，输入 start – all.sh 命令启动 Spark 集群：

```
cd  $ SPARK_HOME/sbin
./start – all.sh
```

（7）Spark 集群启动验证

输入 jps 命令，Master 主机上会显示 master 进程。

```
#jps
5378 NameNode
5608 SecondaryNameNode
7260 Jps
7181 Master
5742 ResourceManager
```

输入 jps 命令，在 worker 结点上会显示 worker 进程。

```
#jps
4152 Worker
3994 NodeManager
4202 Jps
3262 DataNode
```

说明已经成功安装好了 Spark。

1.3.2　安装 MySQL

MySQL 是一款开源的关系型数据库管理系统，有很多版本，MySQL 的安装方式也有很多种，这里介绍的是 Ubuntu14.04 LTS 下 MySQL 5.5 的安装步骤。

Ubuntu 4.04 默认情况下没有安装 MySQL，在安装 MySQL 之前，可以检查系统是否已经安装了 MySQL，如未安装，则只要在联网情况下使用 sudo apt – get install mysql – server 命令即可安装。

在安装之前，可以先检查系统中是否已经安装了 mysql，命令如下：

```
netstat – tap | grep mysql
```

结果如图 1-5 所示，表明目前系统中未安装 mysql。

第1章 认识Spark SQL

```
hadoop@hadoop:~$ netstat -tap | grep mysql
(Not all processes could be identified, non-owned process info
 will not be shown, you would have to be root to see it all.)
```

图 1-5　检查系统是否安装了 MySQL

在 Ubuntu 中安装 MySQL 非常简单，只需要使用如下几条简单的指令即可：
1）更新最新的软件源中的软件列表：

　　sudo apt – get update

2）更新已安装软件到最新版本：

　　sudo apt – get upgrade

3）安装 MySQL，这里默认安装 MySQL 5.5：

　　sudo apt – get install mysql – server

具体的程序运行情况，如图 1-6 所示。

```
hadoop@hadoop:~$ sudo apt-get install mysql-server
Reading package lists... Done
Building dependency tree
Reading state information... Done
The following extra packages will be installed:
  libaio1 libdbd-mysql-perl libdbi-perl libhtml-template-perl libmysqlclient18
  libterm-readkey-perl mysql-client-5.5 mysql-client-core-5.5 mysql-common
  mysql-server-5.5 mysql-server-core-5.5
Suggested packages:
  libmldbm-perl libnet-daemon-perl libplrpc-perl libsql-statement-perl
  libipc-sharedcache-perl tinyca mailx
The following NEW packages will be installed:
  libaio1 libdbd-mysql-perl libdbi-perl libhtml-template-perl libmysqlclient18
  libterm-readkey-perl mysql-client-5.5 mysql-client-core-5.5 mysql-common
  mysql-server mysql-server-5.5 mysql-server-core-5.5
0 upgraded, 12 newly installed, 0 to remove and 0 not upgraded.
Need to get 9,005 kB of archives.
After this operation, 97.1 MB of additional disk space will be used.
Do you want to continue? [Y/n]
```

图 1-6　安装 mysql – server

输入"Y"继续执行安装指令。
4）输入 root 用户的登录密码。
在安装过程中，shell 中会弹出如图 1-7 所示的对话框，提示输入 root 用户的登录密码。
再次输入确定密码后，就配置好了 root 用户的密码。然后安装程序会继续执行安装，直到安装完成。
5）测试 MySQL 是否安装成功。
安装完成后可以测试一下 MySQL 是否安装成功，命令如下：

　　mysql – u root – p <password>
　　<password>（按照提示输入你的密码）

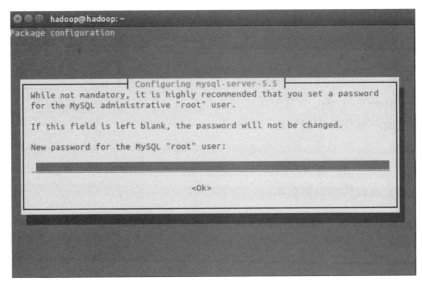

图 1-7　提示输入登录密码用的对话框

登录成功会出现如图 1-8 所示的界面。

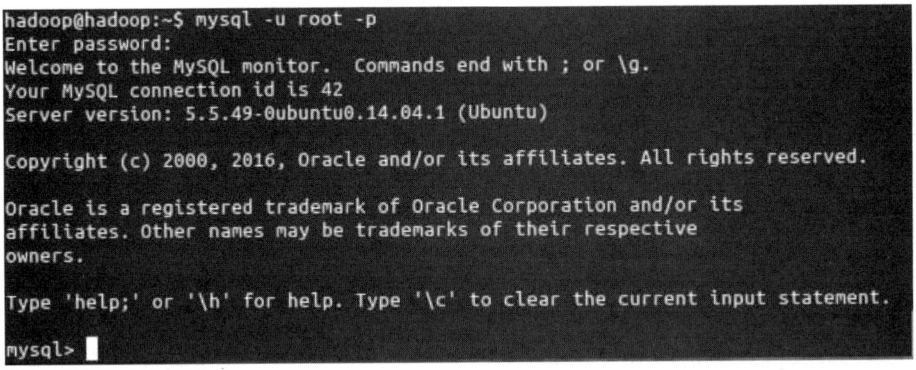

图 1-8　MySQL 登录成功

可以测试一条简单的 SQL 语句，比如：查询 MySQL 默认有几个数据库，如图 1-9 所示：

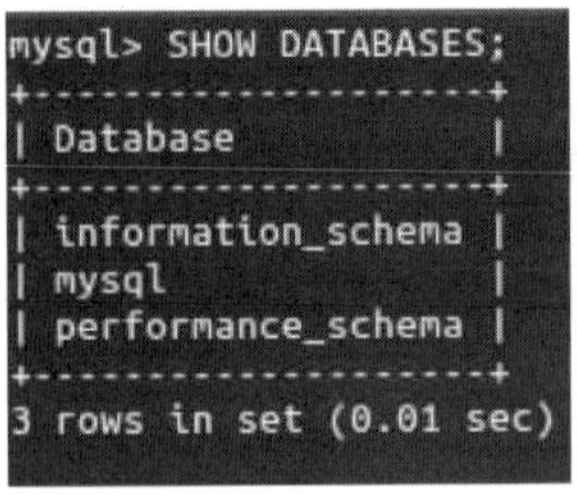

图 1-9　列出默认数据库

至此，MySQL 就安装结束了。

1.3.3 启动 Hive Metastore

使用 Spark SQL，并使用 Hive 作为数据仓库，需要在安装了 Hive 的那台机器上的 Spark 的 conf 目录下，配置 Hive 的元数据信息。这样即使不启动 Hive，Spark 也能正常工作。

首先，进入到 Spark 安装目录下的 conf 目录，执行下面的指令：

```
vim hive-site.xml
```

将如下信息添加到 hive-site.xml 文件中：

```
<?xml version="1.0" encoding="UTF-8"?>
<configuration>
<property>
<name>hive.metastore.uris</name>
<value>thrift://master:9083</value>
</property>
</configuration>
```

- hive.metastore.uris：Hive 连接到该 URL 请求远程元存储的元数据。Spark SQL 通过连接 Hive 提供的 Metastore 服务来获取 Hive 表的元数据。
- URL 对应的值为 thrift://master:9083。

配置好 hive-site.xml，就可以启动 Metastore 服务了，并把它做为后台进程。

```
hive --servicemetastore > metastore.log 2>&1 &
```

Section 1.4 Spark SQL 初试

1.4.1 通过 spark-shell 来使用 Spark SQL

按下面的步骤来开启 Spark-Shell 的使用。

1）启动 Hive 的 Metastore 服务。

```
hive --servicemetastore > metastore.log 2>&1 &
```

2）示例数据准备。

接下来的示例会读取 HDFS 上的 people.json 文件，所以需要预先从本地把该文件 put 到 HDFS 中，该文件在 Spark 的 examples 子目录中。数据准备过程如下：

```
###进入本地 Spark 安装目录下的 examples 子目录中
root@ master:~# cd $SPARK_HOME
root@ master:/opt/software/spark-1.6.3-bin-hadoop2.6# cd examples/src/main/resources

###列出该目录下的文件,发现 people.json 就在该目录中
root@ master:/opt/software/spark-1.6.3-bin-hadoop2.6/examples/src/main/resources# ll
    total 40
    drwxr-xr-x 2 500 500 4096 Jul 20 05:28 ./
    drwxr-xr-x 7 500 500 4096 Jul 20 05:28 ../
    -rw-r--r-- 1 500 500  240 Jul 20 05:28 full_user.avsc
    -rw-r--r-- 1 500 500   73 Jul 20 05:28 people.json
    -rw-r--r-- 1 500 500   32 Jul 20 05:28 people.txt
    -rw-r--r-- 1 500 500  185 Jul 20 05:28 user.avsc
    -rw-r--r-- 1 500 500  334 Jul 20 05:28 users.avro
    -rw-r--r-- 1 500 500  615 Jul 20 05:28 users.parquet

###在 hdfs 上创建/user/root/examples 目录,以便存储示例数据文件
root@ master:~#hdfs dfs -mkdir /user/root
root@ master:~#hdfs dfs -mkdir /user/root/examples

###从本地当前物理目录下把示例文件 people.json 上传到 hdfs 上
root@ master:~#hdfs dfs -put /opt/software/spark-1.6.3-bin-hadoop2.6/examples/src/main/
    resources/people.json    /user/root/examples

###查看 hdfs 指定目录下的文件
root@ master:~#hdfs dfs -ls /user/root/examples
    Found 1 items
    -rw-r--r--   2 root supergroup   73 2016-11-10 12:09 /user/root/examples/people.json
```

3) 启动 spark-shell。

```
spark-shell --master spark://master:7077
```

4) 运行示例。

通过 Spark SQL 读取 HDFS 中的 people.json 文件,并查看操作这个数据文件。具体运行情况如下:

```
###创建一个 sqlContext
val sqlContext = new org.apache.spark.sql.SQLContext(sc)

###读取 json 中的数据并且创建一个 Dataframe
val df = sqlContext.read.json("examples/people.json")

###查看 dataframe 的内容
df.show()
```

###show 结果如下

```
+----+-------+
| age|   name|
+----+-------+
|null|Michael|
|  30|   Andy|
|  19| Justin|
+----+-------+
```

###查看 dataframe 的树形结构
df.printSchema()
// root
// |-- age:long (nullable = true)
// |-- name:string (nullable = true)

###只查看 name 这一列的所有数据,并且显示出来
df.select("name").show()

```
+-------+
|   name|
+-------+
|Michael|
|   Andy|
| Justin|
+-------+
```

###查看 name,和 age+1 的结果,并且 show 出来
df.select(df("name"),df("age")+1).show()

```
+-------+---------+
|   name|(age + 1)|
+-------+---------+
|Michael|     null|
|   Andy|       31|
| Justin|       20|
+-------+---------+
```

###选出年龄大于 21 岁的人,并且显示出来
df.filter(df("age")>21).show()

```
+---+----+
|age|name|
+---+----+
| 30|Andy|
+---+----+
```

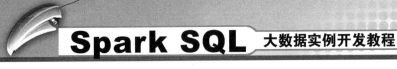

1.4.2　Spark SQL 的命令终端

Spark SQL 的 CLI（命令终端）是一个方便的工具，以本地方式运行在 Hive 的元数据服务上，可以直接在命令行中输入查询语句进行查询。不过需要注意的是，SparkSQL 的 CLI 不能操作 Thrift JDBC Server。下面介绍如何使用 CLI。

首先进入 Spark 安装目录下的 bin 目录，启动 Spark SQL。

```
./spark-sql
```

运行成功后将看到 Spark SQL 命令提示符：

```
spark-sql>
```

接下来进行 Spark SQL CLI 的操作。

1）列出 Hive 中的数据库列表。

在 Spark SQL CLI 中的操作几乎和在 DBMS 中的操作一样，列出 Hive 中有哪些数据库。

```
spark-sql> show databases;
```

结果如下所示（在显示结果之前 CLI 中会打印很多日志信息，这里忽略日志信息，直接显示结果）：

```
default
hive
```

2）选择使用 Hive 数据库。

```
spark-sql> use hive;
```

3）查看数据库中有哪些表，结果中表名后的 false 表示不是临时表。

```
spark-sql> show tables;

//结果如下
sogouq1 false
sogouq2 false
tbdate false
tbstock false
tbstockdetail false
```

4）查看 tbdate 表中有多少条数据。

```
spark-sql> select count(*) from tbdate;
4383
```

第1章 认识Spark SQL

5）查看 tbdate 表的结构。

spark – sql > desc tbdate；

结果如下：

```
dateid       string  NULL
theyearmonth string  NULL
theyear      string  NULL
themonth     string  NULL
thedate      string  NULL
theweek      string  NULL
theweeks     string  NULL
theqout      string  NULL
thetenday    string  NULL
thehalfmonth string  NULL
```

6）查看 tbdate 表的前 10 条数据。

spark – sql > select count(*) from tbdate limit 10；

结果如下：

```
2003 – 1 – 1   200301 2003  1  1   3  1  1  1
2003 – 1 – 2   200301 2003  1  2   4  1  1  1
2003 – 1 – 3   200301 2003  1  3   5  1  1  1
2003 – 1 – 4   200301 2003  1  4   6  1  1  1
2003 – 1 – 5   200301 2003  1  5   7  1  1  1
2003 – 1 – 6   200301 2003  1  6   1  2  1  1
2003 – 1 – 7   200301 2003  1  7   2  2  1  1
2003 – 1 – 8   200301 2003  1  8   3  2  1  1
2003 – 1 – 9   200301 2003  1  9   4  2  1  1
2003 – 1 – 10  200301 2003  1  10  5  2  1  1
```

1.4.3 Spark 的 Web UI

启动 Spark 集群后，即可访问 Spark 的 Web 控制台——Web UI，直接在浏览器中输入 Http://master:8080，或者输入"http://master 机器的 IP 地址:8080"，如图 1-10 所示。

> 预先在 hosts 中插入 Master 和 IP 的映射记录，比如：192.168.1.18 master，这样就可以直接访问 http://master:8080 来打开 Spark 的 Web 控制台。

了解 Spark 的 Web UI 对学习 Spark 非常重要，可以从 Web UI 中非常清楚地了解 Spark 的运行过程，下面简单介绍。

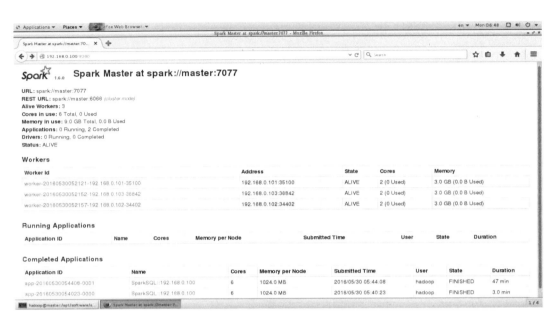

图 1-10　Spark 的 Web 控制台

- 在 Web UI 的最上面是 Spark 集群的地址，集群的基本概况如下：

```
URL:spark://master:7077                              //Spark 的 URL 地址
REST URL:spark://master:6066 (cluster mode)          //Spark 的 REST URL 地址
Alive Workers:3                                      //Spark 集群中活着的 worker 结点数是 3
Cores in use:6 Total,0 Used                          //Spark 集群共 6 个 Cores,使用了 0 个
Memory in use:9.0 GB Total,0.0 B Used                //Spark 集群 9GB 内存,使用了 0 个。
Applications:0 Running,2 Completed                   //Spark 集群的应用 0 个在运行,已经完成 2 个。
Drivers:0 Running,0 Completed                        //Spark 集群 Drivers 0 个运行,0 个完成。
Status:ALIVE                                         //Spark 集群为活着状态
```

从概况中可以知道集群的地址、规模、资源、运行过多少个 APP 及集群的状态。
- 在 Workers 一栏中可以看到每个 worker 的 ID、地址、状态、cores 的个数和内存状态。
- 在 Running Applications 一栏中可以看到目前正在运行的 Applications 的 ID、名称、CPU 资源、每个结点上使用内存资源的情况、程序提交的时间、哪个用户提交的、目前运行的状态、程序运行的时间。
- 在 Completed Applications 这一栏中，是已经运行完成的 Spark 程序的信息。

Section 1.5　本章小结

本章介绍了 Spark SQL 的发展概况，使读者对 Spark SQL 有了初步的了解，讲解了 Spark 环境的搭建、SparkSQL 的基本操作，以及 Hive 的基础操作，并且初步介绍了 Spark SQL 的使用，包括 Spark SQL CLI（命令行界面），以及 Spark Web UI 的基本操作。

第 2 章　DataFrame 原理与常用操作

本章将介绍 DataFrame 编程模型、基本操作、API 的使用，以及 DataFrame 与 RDD 的实战应用。在开始本章 DataFrame 的实践案例之前，需要对 DataFrame 编程模型有一个基础了解。Spark 从 1.3 版本开始引入了新的 DataFrame 编程模型，DataFrame 类似于关系数据库的表，这一组件的引入，简化了 Spark SQL 的处理，并简化了编程复杂度，极大地方便了 Spark SQL 的应用。

2.1 DataFrame 编程模型

DataFrame 这个名字是取自 R、Python 等语言中的概念，DataFrame 在数据统计、分析的语言里扮演了举足轻重的角色。Spark 引入 DataFrame 实现了在大数据平台同样的统计分析功能。随着不断地优化，DataFrame 可以由更广阔的数据源来创建，例如：结构化的数据文件、Hive 表、外部数据库或者现有的 RDD。

在前面 1.1.2 一节中，我们已经了解了 DataFrame 与 RDD 之间的差异。为了更好地理解 DataFrame 究竟是什么，通过图 2-1 再把 DataFrame 与 RDD 进行对比。

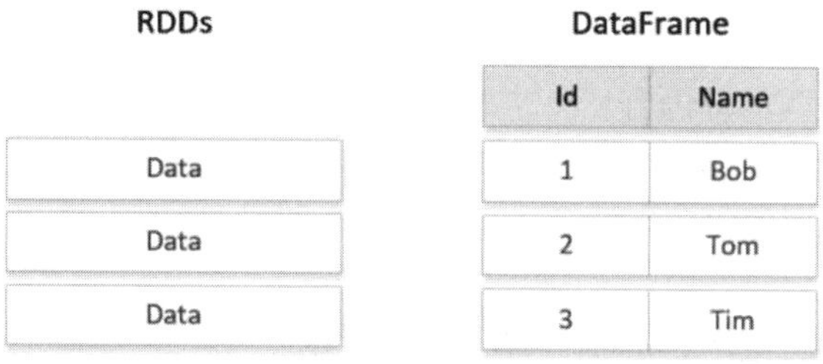

图 2-1　DataFrame 和 RDD 进行对比

从图 2-1 中可以看出：

1）RDD 的每一项数据都是一个整体，这也就导致了 Spark 框架无法洞悉数据记录内部的细节，限制了 Spark SQL 的性能提升。

2）DataFrame 的数据特点如图 2-1 右边所示，其包含了每个数据记录的 Metadata 信息，可以在优化时基于列内部进行优化（例如一共 30 列，如果只需要其中 10 列，那么就可以只获取其中 10 列的信息，而不需要把所有 30 列的数据全部取出）。

DataFrame 更像是 RDD 的加强版，带有更多的细节信息，与普通 RDD 不同的就是 DataFrame 是有 Schema（标记）的，也就是说，DataFrame 是带有每一列信息的。可以把 DataFrame 理解为一个分布式的二维表，每一列都带有名称和类型，这就意味着 Spark SQL 可以基于每一列数据的元数据进行更加细粒度的分析，而不是如同以往分析 RDD 的时候那种粗粒度的分析。于是基于 DataFrame 就可以进行更加高效的性能优化。

DataFrame 编程模型的功能特性如下：
- 从 KB 到 PB 级的数据量支持。
- 多种数据格式和多种存储系统支持。
- 通过 Spark SQL 的 Catalyst 优化器进行优化，生成代码。
- 为 Python、Java、Scala 和 R 语言（SparkR）提供 API。

2.2 DataFrame 基本操作实战

本节给出了一个集团公司对人事信息处理场景的简单案例，详细分析 DataFrame 上的各种常用操作，包括集团子公司间的职工人事信息的合并、职工的部门相关信息查询、职工信息的统计、关联职工与部门信息的统计，以及如何将各种统计得到的信息存储到外部存储系统等。

在本案例中，涉及的 DataFrame 操作包括：
- 从外部文件构建 DataFrame。
- 在 DataFrame 上进行比较常用的操作。
- 多个 DataFrame 之间的操作。
- DataFrame 的持久化操作等。

2.2.1 数据准备

数据准备包含下面两部分内容：
- 创建数据文件。
- 将数据文件上传到 HDFS 存储系统上。

1. 创建数据文件

首先，在本地文件目录中，分别创建下面的数据文件：
- 员工信息：people.json。
- 新增员工：newPeople.json。
- 部门信息：department.json。

1）创建员工信息文件（people.json）。

```
{"name":"Michael","job number":"001","age":33,"gender":"male","depId":1,"salary":3000}
{"name":"Andy","job number":"002","age":30,"gender":"female","depId":2,"salary":4000}
{"name":"Justin","job number":"003","age":19,"gender":"male","depId":3,"salary":5000}
{"name":"John","job number":"004","age":32,"gender":"male","depId":1,"salary":6000}
{"name":"Herry","job number":"005","age":20,"gender":"female","depId":2,"salary":7000}
{"name":"Jack","job number":"006","age":26,"gender":"male","depId":3,"salary":3000}
```

people.json 文件包含了员工的相关信息,每一列分别对应:
员工姓名、工号、年龄、性别、部门 ID 及薪资。
2)创建新增员工信息文件(newPeople.json)。

```
{"name":"John","job number":"007","age":32,"gender":"male","depId":1,"salary":4000}
{"name":"Herry","job number":"008","age":20,"gender":"female","depId":2,"salary":5000}
{"name":"Jack","job number":"009","age":26,"gender":"male","depId":3,"salary":6000}
```

newPeople.json 对应新入职员工的信息,数据结构与员工信息一致。
3)创建部门信息文件(department.json)。

```
{"name":"Development Dept","depId":1}
{"name":"Personnel Dept","depId":2}
{"name":"Testing Department","depId":3}
```

department.json 是部门信息,包含两列:部门名称和部门 ID。
其中,部门 ID 对应员工信息中的部门 ID,即员工信息文件中的 depID 列。

2. 将数据文件上传到 HDFS 存储系统上

1)将这 3 个数据文件上传到 Hadoop 集群中,命令如下:

```
##列出当前目录下的 3 个数据文件
root@ Master:~/test# ls
department.json newPeople.json people.json

###上传用户信息文件到 HDFS
root@ Master:~/test#hadoop dfs -put ./people.json /library/SparkSQL/Data

###上传用户信息文件到 HDFS
root@ Master:~/test#hadoop dfs -put ./newPeople.json /library/SparkSQL/Data

###上传部门信息文件到 HDFS
root@ Master:~/test#hadoop dfs -put ./department.json /library/SparkSQL/Data
```

📖 需要预先在 HDFS 上创建好/library/SparkSQL/Data 文件夹。

2)通过 HDFS 命令查看上传结果。
通过 Hadoop 的命令行,来查看上传文件是否成功。

```
root@ Master:/usr/local/hadoop/hadoop-2.6.0/sbin# hdfs dfs -ls /library/SparkSQL/Data
Found 3 items
-rw-r--r--   3 root supergroup        115 2016-04-23 23:28 /library/SparkSQL/Data/department.json
-rw-r--r--   3 root supergroup        255 2016-04-23 23:27 /library/SparkSQL/Data/newPeople.json
-rw-r--r--   3 root supergroup        514 2016-04-23 23:26 /library/SparkSQL/Data/people.json
```

可以看到，3 个文件已经上传到 HDFS 存储系统上。

3）通过 Hadoop 的 Web 控制台查看上传结果。

还可以通过 Web 控制台的方式查看 HDFS 上的文件，打开网址 http://Master:50070，即可进入 Hadoop 的 Web 控制台，如图 2-2 所示。

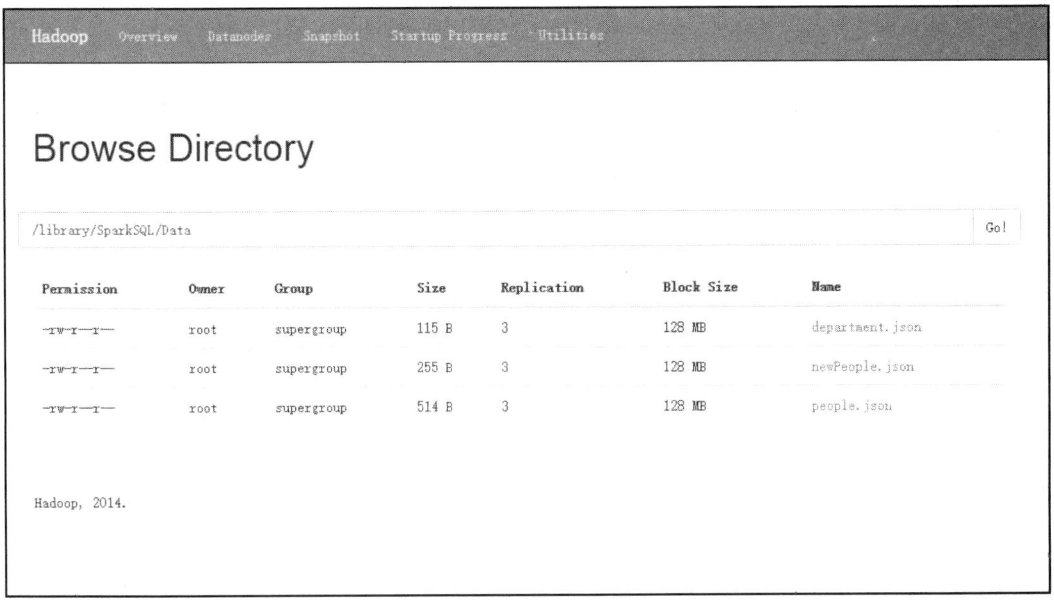

图 2-2　Hadoop Web 控制台

2.2.2　启动交互式界面

按下面步骤启动 Spark – Shell 交互式界面。

1）启动 Spark 集群。

> root@ Master:~# cd /usr/local/spark/spark – 1.6.3 – bin – hadoop2.6/sbin/
> root@ Master:/usr/local/spark/spark – 1.6.3 – bin – hadoop2.6/sbin# ./start – all.sh

2）启动日志管理。

> root@ Master:/usr/local/spark/spark – 1.6.3 – bin – hadoop2.6/sbin# ./start – history – server.sh
> starting org.apache.spark.deploy.history.HistoryServer, logging to /usr/local/spark/spark – 1.6.3 – bin – hadoop2.6/logs/spark – root – org.apache.spark.deploy.history.HistoryServer – 1 – Master.out

3）启动交互式界面。

> root@ Master:/usr/local/spark/spark – 1.6.3 – bin – hadoop2.6/sbin# cd ../bin
> root@ Master:/usr/local/spark/spark – 1.6.3 – bin – hadoop2.6/bin# ./spark – shell – – master spark://Master:7077

第2章 DataFrame原理与常用操作

启动后出现如下信息:

```
...
Welcome to
      ____              __
     / __/__  ___ _____/ /__
    _\ \/ _ \/ _ `/ __/  '_/
   /___/ .__/\_,_/_/ /_/\_\   version 1.6.3
      /_/
...
scala >
```

启动交互式界面完成。

2.2.3 数据处理与分析

这部分内容将对员工信息文件及部门信息文件进行数据处理和分析,下面是具体的操作过程。

1. 修改日志等级

scala > import org. apache. log4j. Level
import org. apache. log4j. Level

scala > import org. apache. log4j. Logger
import org. apache. log4j. Logger

scala > Logger. getLogger("org. apache. spark"). setLevel(Level. WARN)

scala > Logger. getLogger("org. apache. spark. sql"). setLevel(Level. WARN)

将日志等级设置为 Level WARN,是为了简化界面的输出信息。

2. 加载文件

////创建一个 sqlContext
scala > val sqlContext = new org. apache. spark. sql. SQLContext(sc)

scala > val people = sqlContext. jsonFile("hdfs:/library/SparkSQL/Data/people. json")
people: org. apache. spark. sql. DataFrame = [age: bigint, depId: bigint, gender: string, job number: string, name: string, salary: bigint]

scala > val dept   = sqlContext. load("hdfs:/library/SparkSQL/Data/department. json" , "json")
dept: org. apache. spark. sql. DataFrame = [depId: bigint, name: string]

上面示范了两种方式,分别加载 HDFS 上的员工信息文件和部门信息文件,得到了两个 DataFrame 实例: people 和 dept。

3. 以表格形式查看 people 信息

```
scala > people.show
+---+-----+------+-----------+-------+------+
|age|depId|gender|job number | name |salary|
+---+-----+------+-----------+-------+------+
| 33|   1 | male |    001    |Michael| 3000 |
| 30|   2 |female|    002    | Andy  | 4000 |
| 19|   3 | male |    003    |Justin | 5000 |
| 32|   1 | male |    004    | John  | 6000 |
| 20|   2 |female|    005    | Herry | 7000 |
| 26|   3 | male |    006    | Jack  | 3000 |
+---+-----+------+-----------+-------+------+
```

通过 show 方法，可以以表格的形式输出各个 DataFrame 的内容。

4. **DataFrame 基本信息的查询**

````
////查询 people 包含的全部列信息
scala > people.columns
res4:Array[String] = Array(age,depId,gender,job number,name,salary)

////统计 people 包含的记录条数
scala > people.count
res5:Long = 6

////获取前 3 条记录信息，并以数组形式呈现
scala > people.take(3)
res6:Array[org.apache.spark.sql.Row] = Array([33,1,male,001,Michael,3000],[30,2,female,002,Andy,4000],[19,3,male,003,Justin,5000])

////将 people 转换为 JsonRDD，并使用 RDD 的 collect 方法返回
scala > people.toJSON.collect
res7:Array[String] = Array({"age":33,"depId":1,"gender":"male","job number":"001","name":"Michael","salary":3000},{"age":30,"depId":2,"gender":"female","job number":"002","name":"Andy","salary":4000},{"age":19,"depId":3,"gender":"male","job number":"003","name":"Justin","salary":5000},{"age":32,"depId":1,"gender":"male","job number":"004","name":"John","salary":6000},{"age":20,"depId":2,"gender":"female","job number":"005","name":"Herry","salary":7000},{"age":26,"depId":3,"gender":"male","job number":"006","name":"Jack","salary":3000})
````

以上是针对员工信息的 DataFrame 进行一些基本信息的查询操作：

- 使用 DataFrame 的 columns 方法，查询 people 包含的全部列信息，以数组形式返回列名组。

- 使用 DataFrame 的 count 方法，统计 people 包含的记录条数，即员工个数。
- 使用 DataFramed 的 take 方法，获取前 3 条员工记录信息，并以数组形式呈现出来。
- 最后使用 DataFrame 的 toJSON 方法，将 people 转换为 JSONRDD 类型，并使用 RDD 的 collect 方法返回其包含的员工信息。

5. 对员工信息进行条件查询，并输出结果

```
////使用 filter 方法,统计性别为男性的记录数
scala > people.filter("gender = 'male'").count
res8 : Long = 4
//或者使用下面的写法
scala > people.filter( $"gender" === "male" ).count
res8 : Long = 4

////使用 filter 方法,统计性别不为女性的记录数
scala > people.filter( $"gender" !== "female" ).count
res9 : Long = 4
//或者使用下面的写法
scala > people.filter("gender! = 'female'").count
res9 : Long = 4

////使用 filter 方法,查询并显示年龄大于 25 岁的记录
scala > people.filter( $"age" > 25 ).show
+---+-----+------+----------+-------+------+
|age|depId|gender|job number| name  |salary|
+---+-----+------+----------+-------+------+
| 33|    1|  male|       001|Michael|  3000|
| 30|    2|female|       002|   Andy|  4000|
| 32|    1|  male|       004|   John|  6000|
| 26|    3|  male|       006|   Jack|  3000|
+---+-----+------+----------+-------+------+

////使用 where 方法,查询并显示年龄大于 28 岁的记录
scala > people.where( $"age" > 28 ).show
+---+-----+------+----------+-------+------+
|age|depId|gender|job number| name  |salary|
+---+-----+------+----------+-------+------+
| 33|    1|  male|       001|Michael|  3000|
| 30|    2|female|       002|   Andy|  4000|
| 32|    1|  male|       004|   John|  6000|
+---+-----+------+----------+-------+------+

////使用 where 方法,查询并显示年龄大于 25 岁并且性别为男性的记录
```

```
scala > people.where($"age" > 25 && $"gender" === "male").show
+---+-----+------+-----------+-------+------+
|age|depId|gender|job number | name  |salary|
+---+-----+------+-----------+-------+------+
| 33|  1  | male |   001     |Michael| 3000 |
| 32|  1  | male |   004     | John  | 6000 |
| 26|  3  | male |   006     | Jack  | 3000 |
+---+-----+------+-----------+-------+------+
```

////使用 where 方法，查询并显示年龄大于 25 岁的记录
```
scala > people.where('age > 25).show
+---+-----+------+-----------+-------+------+
|age|depId|gender|job number | name  |salary|
+---+-----+------+-----------+-------+------+
| 33|  1  | male |   001     |Michael| 3000 |
| 30|  2  |female|   002     | Andy  | 4000 |
| 32|  1  | male |   004     | John  | 6000 |
| 26|  3  | male |   006     | Jack  | 3000 |
+---+-----+------+-----------+-------+------+
```

在上述示例中，针对员工信息的 DataFrame，进行了一些条件查询操作：
- 使用 count 方法统计了"gender"列为"male"的员工数。
- 基于"age"和"gender"两列，使用不同的查询条件，不同的 DataFrame API，即 where 和 filter 方法，对员工信息进行过滤。
- 最后仍然使用 show 方法，将查询结果以表格的形式呈现出来。
- 在各个例子中，使用了几种不同的方式，作为查询条件的参数。

📖 特别注意上面查询条件表达式中的单引号及 $ 符号。

6. 根据指定的列名，以不同方式进行排序

////先按工号升序排序，再按部门降序排序，显示全部记录
```
scala > people.sort($"job number".asc,col("depId").desc).show
+---+-----+------+-----------+-------+------+
|age|depId|gender|job number | name  |salary|
+---+-----+------+-----------+-------+------+
| 33|  1  | male |   001     |Michael| 3000 |
| 30|  2  |female|   002     | Andy  | 4000 |
| 19|  3  | male |   003     |Justin | 5000 |
| 32|  1  | male |   004     | John  | 6000 |
| 20|  2  |female|   005     | Herry | 7000 |
```

```
|26 |      3 | male   |    006 | Jack    | 3000 |
+---+--------+--------+--------+---------+------+
```

////先按工号升序排序,仅显示前3条记录
scala > people. sort($"job number"). show(3)
//或者
scala > people. sort("job number"). show(3)

```
+---+-----+------+----------+-------+------+
|age|depId|gender|job number| name  |salary|
+---+-----+------+----------+-------+------+
| 33|    1|  male|       001|Michael|  3000|
| 30|    2|female|       002|  Andy |  4000|
| 19|    3|  male|       003| Justin|  5000|
+---+-----+------+----------+-------+------+
```
only showing top 3 rows

////先按工号倒序排序,仅显示前3条记录
scala > people. sort($"job number". desc). show(3)

```
+---+-----+------+----------+-----+------+
|age|depId|gender|job number|name |salary|
+---+-----+------+----------+-----+------+
| 26|    3|  male|       006| Jack|  3000|
| 20|    2|female|       005|Herry|  7000|
| 32|    1|  male|       004| John|  6000|
+---+-----+------+----------+-----+------+
```
only showing top 3 rows

在上述示例中,针对员工信息DataFrame,基于"job number"和"depId"两列,使用sort方法,以不同方式进行排序,并输出结果,具体包含:

- 先以"job number"列升序,然后再按"depId"列降序的方式,对people进行排序,并输出排序后的内容;这里给出了两种指定列的方式。
- 以"job number"列进行默认排序(升序),并显示排序后的前3条记录。
- 以"job number"列指定降序方式排序,并显示排序后的前3条记录。

7. 为员工信息增加一列:等级("level")

```
scala > people. withColumn("level",people("age")/10). show
+---+-----+------+----------+-------+------+-----+
|age|depId|gender|job number| name  |salary|level|
+---+-----+------+----------+-------+------+-----+
| 33|    1|  male|       001|Michael|  3000|  3.3|
| 30|    2|female|       002|  Andy |  4000|  3.0|
```

```
| 19 |  3 | male   |  003 | Justin | 5000 | 1.9 |
| 32 |  1 | male   |  004 | John   | 6000 | 3.2 |
| 20 |  2 | female |  005 | Herry  | 7000 | 2.0 |
| 26 |  3 | male   |  006 | Jack   | 3000 | 2.6 |
+----+----+--------+------+--------+------+-----+
```

在上述示例中，通过 withColumns 方法增加了新的一列等级信息，列名为"level"。在 withColumns 方法中：

第一个参数"level"指定了新增列的列名。

第二个参数 people("age")/10，指定了该列的实例，通过转换得到新列，people("age")调用了 DataFrame 的 apply 方法，返回"age"列名所对应的列。

8. 修改工号列名

```
scala > people.columns
res13: Array[String] = Array(age,depId,gender,job number,name,salary)

scala > people.withColumnRenamed("job number","jobId").columns
res15: Array[String] = Array(age,depId,gender,jobId,name,salary)
```

在上述示例中，通过 withColumnRenamed 方法修改列名，示例将 people 的"job number"列名修改为"jobId"，通过交互式输出信息可以看到列名已经被修改。

注意，修改的列名如果不存在，不会报错，但列名不会修改，如下所示：

```
scala > val rnjobnum = people.withColumnRenamed("job numbe","jobId")
rnjobnum: org.apache.spark.sql.DataFrame = [age: bigint, depId: bigint, gender: string, job number: string, name: string, salary: bigint]

scala > rnjobnum.columns
res16: Array[String] = Array(age,depId,gender,job number,name,salary)

scala > people.columns
res17: Array[String] = Array(age,depId,gender,job number,name,salary)
```

在该示例中，指定修改的"job numbe"列名拼写错误（少了一个字母 r），所以正确的列名"job number"并没有修改成功。

9. 增加新员工

```
////使用 jsonFile 方法加载新员工信息文件
scala > val newPeople = sqlContext.jsonFile("hdfs:/library/SparkSQL/Data/newPeople.json")
newPeople: org.apache.spark.sql.DataFrame = [age: bigint, depId: bigint, gender: string, job number: string, name: string, salary: bigint]

////展示新员工信息
```

```
scala > newPeople.show
+---+-----+------+----------+------+------+
|age|depId|gender|job number| name |salary|
+---+-----+------+----------+------+------+
| 32|   1 | male |   007    | John | 4000 |
| 20|   2 |female|   008    |Herry | 5000 |
| 26|   3 | male |   009    | Jack | 6000 |
+---+-----+------+----------+------+------+

////合并 people 和 newPeople
scala > people.unionAll(newPeople).show
+---+-----+------+----------+--------+------+
|age|depId|gender|job number|  name  |salary|
+---+-----+------+----------+--------+------+
| 33|   1 | male |   001    |Michael | 3000 |
| 30|   2 |female|   002    | Andy   | 4000 |
| 19|   3 | male |   003    |Justin  | 5000 |
| 32|   1 | male |   004    | John   | 6000 |
| 20|   2 |female|   005    | Herry  | 7000 |
| 26|   3 | male |   006    | Jack   | 3000 |
| 32|   1 | male |   007    | John   | 4000 |
| 20|   2 |female|   008    | Herry  | 5000 |
| 26|   3 | male |   009    | Jack   | 6000 |
+---+-----+------+----------+--------+------+
```

在上述示例中,使用 jsonFile 方法加载了新员工信息的文件,然后调用 people 的 unionAll 方法,将新加载的 newPeople 合并进来。

注意:因为加载文件是 lazy 性质的,由于没有对 DataFrame 进行缓存,因此最终合并时会重新加载新旧两个员工信息文件。

10. 查同名员工

```
////通过 unionAll 方法将 people 和 new People 两个信息文件进行合并
////然后使用 groupBy 方法将合并后的 DataFrame 按照"name"列进行分组统计
scala > val groupName = people.unionAll(newPeople).groupBy(col("name")).count
groupName:org.apache.spark.sql.DataFrame = [name:string,count:bigint]

scala > groupName.show
+-------+-----+
| name  |count|
+-------+-----+
```

```
|  Jack   |  2  |
|  John   |  2  |
|  Andy   |  1  |
| Michael |  1  |
| Justin  |  1  |
| Herry   |  2  |
+---------+-----+
```

////使用 filter 方法,过滤"count"列大于 1 的记录并显示
scala > groupName.filter($"count" >1).show

```
+------+-----+
| name | count|
+------+-----+
| Jack |  2  |
| John |  2  |
| Herry|  2  |
+------+-----+
```

////使用函数式编程范式对前两个合并进行分组统计并显示结果
scala > people.unionAll(newPeople).groupBy(col("name")).count.filter($"count" <2).show

```
+---------+-----+
|  name   |count|
+---------+-----+
|  Andy   |  1  |
| Michael |  1  |
| Justin  |  1  |
+---------+-----+
```

在上述示例中,首先通过 unionAll 方法将 people 和 newPeople 两个文件进行合并,然后使用 groupBy 方法将合并后的 DataFrame 按照"name"列进行分组,得到 GroupData 类的实例,实例会自动带上分组的列,以及"count"列。

GroupData 类提供了一组非常有用的统计操作,这里调用它的 count 方法,最终实现对员工名字的分组统计。

📖 GroupData 类在 Spark 2.0.x 版本改为 RelationalGroupedDataset。

11. 分组统计信息

////调用 groupBy 方法得到 GroupData 实例,再调用 agg 方法
scala > val depAgg = people.groupBy("depId").agg(Map(
    | "age" -> "max",

```
            |   "gender" -> "count"
            | ))
depAgg:org.apache.spark.sql.DataFrame = [depId:bigint,max(age):bigint,count(gender):bigint]

scala > depAgg.show
+-----+--------+-------------+
|depId|max(age)|count(gender)|
+-----+--------+-------------+
|    1|      33|            2|
|    2|      30|            2|
|    3|      26|            2|
+-----+--------+-------------+
```

////调用 DataFrame 的 toDF 方法,重新命名 depAgg 的全部列名,增加列名的可读性
```
scala > depAgg.toDF("depId","maxAge","countGender").show
+-----+------+-----------+
|depId|maxAge|countGender|
+-----+------+-----------+
|    1|    33|          2|
|    2|    30|          2|
|    3|    26|          2|
+-----+------+-----------+
```

在上述示例中,首先针对 people 的 "depId" 进行分组,再对分组后得到的 GroupData 实例继续调用 agg 方法,分别对 "age" 列求最大值,对 "gender" 进行分组统计,返回 DataFrame 对象实例 depAgg。depAgg 的 schema 为[depId:bigint,max(age):bigint,count(gender):bigint],即除了带上分组用的 "depId" 列外,还带上列聚合操作后的两列信息。

12. 名字去重

////通过 select 选取 name 列并显示
```
scala > people.unionAll(newPeople).select("name").show
+-------+
|   name|
+-------+
|Michael|
|   Andy|
| Justin|
|   John|
|  Herry|
|   Jack|
```

```
|  John  |
|  Herry |
|  Jack  |
+--------+
```

////通过 select 选取 name 列,在通过 distinct 方法去重之后显示
scala > people.unionAll(newPeople).select("name").distinct.show

```
+--------+
|  name  |
+--------+
|  Jack  |
|  John  |
|  Andy  |
| Michael|
|  Justin|
|  Herry |
+--------+
```

在上述示例中,首先显示新旧员工信息合并后的"name"列,作为后续去重的比较对象。通过 unionAll 新旧员工信息,并只选择其中的"name"列信息后,出现的"name"信息就出现列重复,通过继续调用 DataFrame 的 distinct 去重方法后,可以去除重复的记录数据。

13. 对比新旧员工表

////显示在 people 中,但是不在 newPeople 中的姓名
scala > people.select("name").except(newPeople.select($"name")).show

```
+--------+
|  name  |
+--------+
|  Andy  |
|  Justin|
| Michael|
+--------+
```

////显示既在 people 中,又在 newPeople 中的姓名
scala > people.select("name").intersect(newPeople.select($"name")).show
16/04/24 14:16:04 INFOmapred.FileInputFormat:Total input paths to process :1
16/04/24 14:16:04 INFOmapred.FileInputFormat:Total input paths to process :1

```
+------+
| name |
+------+
```

第2章 DataFrame原理与常用操作

```
| Herry |
| John  |
| Jack  |
+-------+
```

在上述示例中，包含了对 people 和 newPeople 两个员工信息文件中"name"列的两种比较方式，具体如下：

第一种：分别选取 people 和 newPeople 两个员工信息文件中的"name"列，然后通过调用 except 方法，获取在 people 中出现但不在 newPeople 中出现的"name"信息，最后以表格形式呈现结果。

第二种：求"name"的交集，即分别选取 people 和 newPeople 两个员工信息文件中的"name"列，然后通过调用 intersect 方法，获取在 people 中出现但同时又在 newPeople 中出现的"name"信息，最后以表格形式呈现结果。

14. 关联两个 DataFrame 实例

本实例查询员工信息及员工所属的部门：员工信息 people 的 DataFrame 中包括年龄、部门 ID、性别、工号、姓名、薪酬等信息；部门信息 dept 的 DataFrame 中包括部门 ID、部门名称等信息；员工信息 people 和部门信息 dept 根据部门 ID 号进行关联。

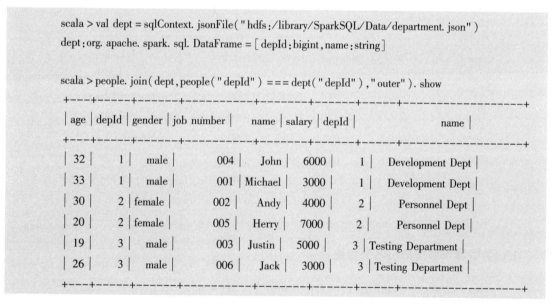

```
scala > val dept = sqlContext.jsonFile("hdfs:/library/SparkSQL/Data/department.json")
dept:org.apache.spark.sql.DataFrame = [depId:bigint, name:string]

scala > people.join(dept,people("depId") === dept("depId"),"outer").show
+---+-----+------+----------+-------+------+-----+------------------+
|age|depId|gender|job number| name  |salary|depId|       name       |
+---+-----+------+----------+-------+------+-----+------------------+
| 32|   1 | male |    004   | John  | 6000 |  1  | Development Dept |
| 33|   1 | male |    001   |Michael| 3000 |  1  | Development Dept |
| 30|   2 |female|    002   | Andy  | 4000 |  2  |  Personnel Dept  |
| 20|   2 |female|    005   | Herry | 7000 |  2  |  Personnel Dept  |
| 19|   3 | male |    003   | Justin| 5000 |  3  | Testing Department|
| 26|   3 | male |    006   | Jack  | 3000 |  3  | Testing Department|
+---+-----+------+----------+-------+------+-----+------------------+
```

在上述示例中，通过调用 join 方法，把 people 中的"depId"列与 dept 中的"depId"列进行 outer join 关联操作。

📖 DataFrame 实例间的 join 关联操作，包括 inner、outer、left outer、right outer、left semi。
- **inner join**：等值连接，只返回两个表中联结字段相等的行。
- **outer join**：包含左、右两个表的全部行，不管另外一边的表中是否存在与它们匹配的行。
- **left_outer join**：如果右边有多行和左边表对应，就每一行都映射输出；如果右边没有行与左边行对应，就输出左边行，右边表字段为 NULL。

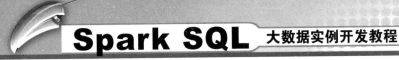

right_outer join：如果左边有多行和右边表对应，就每一行都映射输出；如果左边没有行与右边行对应，就输出右边行，左边表字段为 NULL。

leftsemi join：相当于 SQL 的 in 语句，如果右边有多行和左边表对应，重复的多条记录不输出，只输出一条记录。如果右边没有行与左边行对应，不输出记录。

由于 people 与 dept 的两个 DataFrame 中用于关联的列名相同，都是"depId"，因此，指定关联条件表达式时，需要指出列所属的具体 DataFrame 实例，否则会报错。但是，如果两个列名不同，则可以直接使用列名，表达式会更加精简，比如：

```
////在 dept 中，把 depId 列名重命名为 id,然后赋予新的 rnDept 对象
scala > val rnDept = dept.withColumnRenamed("depId","id")
rnDept:org.apache.spark.sql.DataFrame = [id:bigint,name:string]

////对 people 和 rnDept 进行关联操作，因为列名不同，所以表达式可以直接使用列名
scala > val joinP = people.join(rnDept, $"depId" === $"id","outer")
joinP:org.apache.spark.sql.DataFrame = [age:bigint,depId:bigint,gender:string,job number:string,
name:string,salary:bigint,id:bigint,name:string]

////显示结果
scala > joinP.show
+---+-----+------+----------+------+------+---+------------------+
|age|depId|gender|job number|  name|salary| id|              name|
+---+-----+------+----------+------+------+---+------------------+
| 33|    1|  male|       001|Michael|  3000|  1|  Development Dept|
| 32|    1|  male|       004|  John|  6000|  1|  Development Dept|
| 30|    2|female|       002|  Andy|  4000|  2|    Personnel Dept|
| 20|    2|female|       005| Herry|  7000|  2|    Personnel Dept|
| 26|    3|  male|       006|  Jack|  3000|  3| Testing Department|
| 19|    3|  male|       003|Justin|  5000|  3| Testing Department|
+---+-----+------+----------+------+------+---+------------------+
```

15. 关联操作后按部门名分组统计

```
scala > val joinGp = joinP.groupBy(dept("name")).agg(Map(
        "age"    -> "max",
        "gender" -> "count"
        ))
joinGp:org.apache.spark.sql.DataFrame = [name:string,max(age):bigint,count(gender):bigint]

scala > joinGp.show
+----------------+--------+-------------+
|            name|max(age)|count(gender)|
```

第2章 DataFrame原理与常用操作

```
+------------------+----------+----------------+
|   Personnel Dept |    30    |       2        |
|  Development Dept|    33    |       2        |
| Testing Department|   26    |       2        |
+------------------+----------+----------------+
```

在上述示例中，joinP 是关联操作示例中 people 与 rnDept 进行 join 的结果，对 joinP 对象调用其 groupBy 方法，根据 dept 的 "name" 列进行分组，并在分组后对指定的列执行指定的聚合操作，这里对 "age" 列求最大值，对 "count" 列进行计数。joinGP 是 joinP 根据部门名称进行分组，然后计算最大年龄和性别计数以后生成的 DataFrame。

16. 保存为表

在对各个 DataFrame 实例进行操作后，获取了目标信息，如果后续需要这些信息的话，就必须执行持久化操作，即将文件保存到存储系统或表中。

下面给出几种持久化的示例。

1）首先，将实例持久化到表中。

```
scala > people.saveAsTable("peopletable")

scala > val rnpeople = people.withColumnRenamed("job number","jobId")
rnpeople: org.apache.spark.sql.DataFrame = [age: bigint, depId: bigint, gender: string, jobId: string,
    name: string, salary: bigint]

scala > rnpeople.saveAsTable("rnpeopletable")
```

DataFrame 相关的 Save 操作还有 registerTempTable。

> 在 Spark 2.0.X 版本中，DataFrame 类下没有 saveAsTable 方法，该方法被放在 DataFrameWriter 类下。

2）保存为 JSON 文件。

```
scala > people.save("hdfs:/library/SparkSQL/Data/peoplesave.json","json")
```

这里使用 save 方法，在方法中指定数据源格式为 "json"，可以将 DataFrame 实例持久化到指定的路径上。通过 HadoopWeb Interface 界面可以查看到 JSON 文件。

3）保存为 parquet 文件。

```
scala > people.save("library/SparkSQL/Data/hsqlDF.parquet","parquet")
```

这里同样使用 save 方法，在方法中指定数据源格式为 "parquet"，可以将 DataFrame 实例持久化到指定的路径上。

2.3 通过 RDD 来构建 DataFrame

DataFrame 可以从结构化文件、Hive 表、外部数据库及 RDD 加载构建得到。具体的结构化文件、Hive 表、外部数据库的相关加载可以参考其他章节，这里主要介绍如何通过 RDD 来构建 DataFrame。

Spark SQL 支持两种方式将存在的 RDD 转化为 DataFrame。

第一种：使用反射来推断包含特定对象类型的 RDD 的模式（Schema）。适用对已知数据结构的 RDD 转换，基于反射的方式，代码比较简洁。

第二种：是通过一个编程接口来实现的，这个接口允许构造一个模式，然后在存在的 RDD 上使用它。虽然这种方法代码较为冗长，但是它允许在运行期间不知道列及列类型的情况下构造 DataFrame。

下面对这两种方式做进一步解释。

1. 利用反射推断模式

Spark SQL 能够将含 Row 对象的 RDD 转换成 DataFrame，并推断数据类型。通过将一个键值对（key/value）列表作为 kwargs 传给 Row 类来构造 Rows。key 定义了表的列名，类型通过第一列数据来推断（所以这里 RDD 的第一列数据不能有缺失）。未来版本中将会通过看更多数据来推断数据类型，就像现在对 JSON 文件的处理一样。

2. 编程指定模式

通过编程指定模式需要 3 步：

1）从原来的 RDD 创建一个元组或列表的 RDD。

2）用 StructType 创建一个和步骤 1）中创建的 RDD 中元组或列表的结构相匹配的模式。

3）通过 SQLContext 提供的 createDataFrame 方法将模式应用到 RDD 上。

动态构造有时候有些麻烦，但是 Spark 已经提供了一个 API，就是 DataSet，DataSet 可以基于 RDD，RDD 里面有类型，这样就可以实现 RDD 与 DataFrame 的转换。

RDD + DataFrame + DataSet 最终会形成三足鼎立的局面。从 Spark 2.0 开始，会大量使用 DataSet，因为 DataSet 上可以直接查询，操作起来也会非常直观方便。DataSet 的底层是钨丝计划，提供了更优的性能改进，DataSet 的目标是要所有的子框架都用 DataSet 来进行计算。

接下来示范如何通过编程接口来实现 RDD 与 DataFrame 的转换。

1）首先准备数据

准备的数据文件如下，是名为 persons.txt 的文本文件。

1001,张三,18
1002,李四,19
……
1098,王五,28
1099,赵六,20

第2章 DataFrame原理与常用操作

数据格式如下：
- 第一列为 ID。
- 第二列是姓名。
- 第三列是年龄。

2）使用 Java 实践 RDD 与 DataFrame 的转换。

```java
public class RDD2DataFrameByProgrammatically {
    public static void main(String[] args) {
        SparkConf conf = new SparkConf().setMaster("local")
                .setAppName("RDD2DataFrameByProgrammatically");
        JavaSparkContext sc = new JavaSparkContext(conf);
        SQLContext sqlContext = new SQLContext(sc);
        JavaRDD<String> lines = sc.textFile("E://persons.txt");

        /** * 第一步:在 RDD 的基础上创建类型为 Row 的 RDD */
        JavaRDD<Row> personsRDD = lines.map(new Function<String,Row>() {
            @Override
            public Row call(String line) throws Exception {
                String[] splited = line.split(",");
                return
        RowFactory.create(Integer.valueOf(splited[0]),splited[1],Integer.valueOf(splited[2]));
            }
        });

        /*** 第二步:动态构造 DataFrame 的元数据,一般而言,有多少列及每列的具体类型可能
        来自于 JSON 文件,也可能来自于 DB */
        List<StructField> structFields = new ArrayList<StructField>();
        structFields.add(DataTypes.createStructField("id",DataTypes.IntegerType,true));
        structFields.add(DataTypes.createStructField("name",DataTypes.StringType,true));
        structFields.add(DataTypes.createStructField("age",DataTypes.IntegerType,true));

        //构建 StructType,用于最后 DataFrame 元数据的描述
        StructType structType = DataTypes.createStructType(structFields);

        /*** 第三步:基于 MetaData 及 RDD<Row> 来构造 DataFrame */
        DataFrame personsDF = sqlContext.createDataFrame(personsRDD,structType);

        /** 第四步:注册成为临时表以供后续的 SQL 查询操作 */
        personsDF.registerTempTable("persons");

        /** 第五步,进行数据的多维度分析 */
        DataFrame result = sqlContext.sql("select * from persons where age > 20");
```

45

```java
/**第六步:对结果进行处理,包括由 DataFrame 转换成为 RDD<Row>,以及结果持久化*/
List<Row> listRow = result.javaRDD().collect();
for(Row row:listRow){
System.out.println(row);
}
}
}
```

> 在 Spark SQL 中,建议使用 HiveContext 来替代 SQLContext,它比 SQLContext 功能更强

以上 Java 代码实现了 RDD 与 DataFrame 的转换,非常简单,主要涉及以下步骤:
- 读取文本文件,遍历全部行,创建 RDD<Row>。
- 构造 DataFrame 的元数据(MetaData)。
- 基于 MetaData、RDD<Row>来构造 DataFrame。
- 注册成为临时表。
- 使用上面的临时表,进行数据查询。
- 对结果进行处理:DataFrame 转换成为 RDD<Row>或者持久化。

3) 使用 Scala 实践 RDD 与 DataFrame 的转换。

```scala
import org.apache.spark.{SparkContext,SparkConf}
import org.apache.spark.sql.SQLContext

class RDD2DataFrameByProgrammaticallyScala{
  def main(args:Array[String]):Unit = {
    val conf = newSparkConf()
    conf.setAppName("RDD2DataFrameByProgrammaticallyScala")//设置应用程序的名称
    conf.setMaster("local")

    val sc = newSparkContext(conf)
    val sqlContext = new SQLContext(sc)

    val people = sc.textFile("E://persons.txt")
    val schemaString = "name age"

    import org.apache.spark.sql.Row;
    import org.apache.spark.sql.types.{StructType,StructField,StringType};

    val schema =
   StructType(schemaString.split(" ").map(fieldName => StructField(fieldName,StringType,true)))

    val rowRDD = people.map(_.split(",")).map(p => Row(p(0),p(1).trim))
```

```
    val peopleDataFrame = sqlContext.createDataFrame(rowRDD,schema)
peopleDataFrame.registerTempTable("people")
    val results = sqlContext.sql("select name from people")
    results.map(t => "Name:" + t(0)).collect().foreach(println)
  }
}
```

以上通过 Scala 代码实现了上面 Java 代码一样的效果，具体实现步骤也基本一致。

Section 2.4 缓存表（列式存储）

Spark SQL 可以通过调用 sqlContext.cacheTable("tableName")方法来将一张表缓存到内存中，极大地提高查询效率。然后，Spark 将会仅仅浏览需要的列并且自动地压缩数据，以减少内存的使用及垃圾回收的压力。可以通过调用 uncacheTable("tableName")方法在内存中删除表。

注意： 如果调用 schemaRDD.cache()而不是 sqlContext.cacheTable(...)，将不会用列式格式来缓存（即列式存储），强烈推荐调用 sqlContext.cacheTable(...)。

可以在 SQLContext 上使用 setConf 方法或者在用 SQL 时运行"SET key = value"命令来配置内存缓存，部分配置信息如表 2-1 所示。

表 2-1 用"SET key = value"命令的部分属性配置

属性名称	默认值	含义
Spark.sql.inMemoryColummarStorage.compressed	true	当设置为 true 时，Spark SQL 将为基于数据统计信息的每列自动选择一个压缩算法
Spark.sql.inMemoryColummarStorage.BatchSiz	10000	列式缓存的批数据大小。更大的批数据可以提高内存的利用率以压缩效率，但有 OOM 的风险

下面是对表进行缓存的示例。

```
////从 HDFS 读取 people.json
scala > val people = sqlContext.jsonFile("hdfs:/library/SparkSQL/Data/people.json")
people: org.apache.spark.sql.DataFrame = [age: bigint, depId: bigint, gender: string, job number:
    string, name: string, salary: bigint]

////注册临时表
scala > people.registerTempTable("people")

////访问该临时表
scala > val results = sqlContext.sql("SELECT name FROM people")
results: org.apache.spark.sql.DataFrame = [name: string]

////显示结果
```

```
scala > results.show
+-------+
|  name |
+-------+
|Michael|
|  Andy |
| Justin|
|  John |
|  Herry|
|  Jack |
+-------+
```

////把 people 这张临时表缓存到内存中
```
scala > sqlContext.cacheTable("people")
```

////再次执行 show,执行效率比第一次高很多
```
scala > results.show
+-------+
|  name |
+-------+
|Michael|
|  Andy |
| Justin|
|  John |
|  Herry|
|  Jack |
+-------+
```

////调用 uncacheTable 释放缓存,可以查看 Web Interface 界面,内存会马上释放
```
scala > sqlContext.uncacheTable("people")
```

📖 CacheTable 操作是 lazy 级别的,在调用该方法后不会立即执行。

2.5 DataFrame API 应用示例

　　DataFrame 提供了一套丰富的 API,让 Spark 变得更加平易近人,使得大数据分析的开发越来越容易。DataFrame API 将关系型的处理与过程型处理结合起来,可以对外部数据源(Hive、JSON 等)和 Spark 内建的分布式集合(RDD)进行关系型操作。

第2章 DataFrame原理与常用操作

DataFrame 能处理的外部数据源,除了内置的 Hive、JSON、Parquet、JDBC 以外,还包括 CSV、Avro、HBase 等多种数据源,Spark SQL 多元一体的结构化数据处理能力正在逐渐释放。DataFrame 数据采用压缩的列式存储,对 DataFrame 的操作采用 Catalyst———一种关系操作优化器(也称为查询优化器),因此效率更高。

本节读取电影票房收入的数据文件,生成名为 film 的 DataFrame,对读 DataFrame 应用各种 Spark 算子进行计算。

首先准备数据。模拟生成电影票房收入的数据文件 film.json、newFilm.json,从本地上传到 Hdfs 文件系统中,电影票房收入文件的格式包括6列:票房收入、制片地区、影片ID、语言代码、影片名、上映年份。电影票房收入的数据文件内容如下:

```
{"boxoffice":33.9,"country":"CN","id":"1","languageid":1,"name":"Mermaid","year":2016}
{"boxoffice":24.38,"country":"CN","id":"2","languageid":1,"name":"MonsterHunt","year":2015}
{"boxoffice":13.82,"country":"USA","id":"3","languageid":2,"name":"Avatar","year":2010}
{"boxoffice":15.3,"country":"USA","id":"4","languageid":2,"name":"Zootropolis","year":2016}
{"boxoffice":5.33,"country":"Uk","id":"5","languageid":2,"name":"Spectre","year":2015}
{"boxoffice":3.82,"country":"UK","id":"6","languageid":2,"name":"Sherlock","year":2016}
{"boxoffice":1.54,"country":"TH","id":"7","languageid":3,"name":"FirstLove","year":2012}
```

模拟生成语言代码文件 language.json,同样从本地上传到 Hdfs 文件系统中,语言代码文件的格式包括2列:语言名称、语言代码。语言代码文件的数据内容如下:

```
{"language":"chinese","languageid":1}
{"language":"english","languageid":2}
{"language":"other","languageid":3}
```

接下来在 Spark – Shell 中使用 sqlContext.jsonFile 方法分别加载电影票房收入文件,语言代码文件。

```
scala > val newfilm = sqlContext.jsonFile("hdfs:/library/SparkSQL/Data/newFilm.json")
newfilm: org.apache.spark.sql.DataFrame = [boxoffice:double, country:string, id:string, languageid:
    bigint, name:string, year:bigint]

////语言代码数据。数据内容格式有两列:语言名称和语言代码
scala > val language = sqlContext.jsonFile("hdfs:/library/SparkSQL/Data/language.json")
language: org.apache.spark.sql.DataFrame = [language:string, languageid:bigint]

////把 film 对象赋值给 df,film 是 DataFrame 类型,赋值变量 df
```

49

```
scala > val df = film
df:org.apache.spark.sql.DataFrame = [boxoffice:double, country:string, id:string, languageid:bigint,
       name:string, year:bigint]

////把 newfilm 对象赋值给 newDf,newDf 是 DataFrame 类型,赋值变量 newDf
scala > val newDf = newfilm

////创建一个 sqlContext
scala > val sqlContext = new org.apache.spark.sql.SQLContext(sc)
```

接下来对常用的 DataFrame API 进行示例解析,相应的操作数据为电影票房收入的数据文件 film.json 和 newFilm.json。

1. collect 与 collectAsList

1) 定义。

```
def collect():Array[Row]
def collectAsList():List[Row]
```

2) 功能描述。
- collect 返回一个数组,包含 DataFrame 中包含的全部数据记录。
- collectAsList 返回一个 Java List,包含 DataFrame 中包含的全部数据记录。

3) 示例。

```
scala > val df = film

////show 出 DataFrame 中的数据,以表格形式呈现其内容
scala > df.show
+--------+-------+---+----------+------------+----+
|boxoffice|country| id|languageid|        name|year|
+--------+-------+---+----------+------------+----+
|    33.9|     CN|001|         1|     Mermaid|2016|
|   24.38|     CN|002|         1|Monster Hunt|2015|
|   13.82|    USA|003|         2|      Avatar|2010|
|    15.3|    USA|004|         2|  Zootropolis|2016|
|    5.33|     Uk|005|         2|     Spectre|2015|
|    3.82|     UK|006|         2|    Sherlock|2016|
|    1.54|     TH|007|         3|  First Love|2012|
+--------+-------+---+----------+------------+----+

//// collect 方法返回一个数组,包含 DataFrame 中的全部 Rows
scala > df.collect
```

第2章 DataFrame原理与常用操作

```
res18:Array[org.apache.spark.sql.Row] = Array([33.9,CN,001,1,Mermaid,2016],[24.38,CN,
    002,1,Monster Hunt,2015],[13.82,USA,003,2,Avatar,2010],[15.3,USA,004,2,Zootropolis,
    2016],[5.33,Uk,005,2,Spectre,2015],[3.82,UK,006,2,Sherlock,2016],[1.54,TH,007,3,
    First Love,2012])

////collectAsList 方法返回一个 Java List
scala > df.collectAsList
res19:java.util.List[org.apache.spark.sql.Row] = [[33.9,CN,001,1,Mermaid,2016],[24.38,
    CN,002,1,Monster Hunt,2015],[13.82,USA,003,2,Avatar,2010],[15.3,USA,004,2,
    Zootropolis,2016],[5.33,Uk,005,2,Spectre,2015],[3.82,UK,006,2,Sherlock,2016],
    [1.54,TH,007,3,First Love,2012]]
```

4）解析。

上述示例中首先加载 film，然后以表格的形式呈现其内容。

两个 collect 型的方法都可以获取 df 的全部数据记录，只是返回的类型不同。调用 collect 方法，返回的是数组，数组元素的类型为 org.apache.spark.sql.Row；调用 collectASList 返回类型 java.util.List。

2. count

1）定义。

```
def count():long
```

2）功能描述。

返回 DataFrame 的数据记录的条数。

3）示例。

```
scala > df.count
16/04/27 14:33:45 INFOmapred.FileInputFormat:Total input paths to process :1
res20:Long = 7
```

3. describe

1）定义。

```
def describe(cols:String * ):DataFrame
```

2）功能描述。

概要与描述性统计（Summary and Descriptive Statistics），包含计数、平均值、标准差、最大值和最小值运算。

3）示例。

```
scala > df.describe("name").show
    +--------+------------+
```

```
| summary |       name |
+---------+------------+
|   count |          7 |
|    mean |       null |
|  stddev |       null |
|     min |     Avatar |
|     max | Zootropolis|
+---------+------------+
```

4. First

1）定义。

```
def first( ):Row
```

2）功能描述。

返回 DataFrame 的第一行，等同于 head( )方法。

3）示例。

```
scala > df. first
res22:org. apache. spark. sql. Row = [33.9,CN,001,1,Mermaid,2016]
```

5. head

1）定义。

```
def head( ):Row
```

2）功能描述。

不带参数的 head 方法，返回 DataFrame 的第一条数据记录；指定参数 n 时，则返回前 n 条数据记录。

3）示例。

```
scala > df. head
res23:org. apache. spark. sql. Row = [33.9,CN,001,1,Mermaid,2016]

scala > df. head(3)
res24:Array[ org. apache. spark. sql. Row] = Array([33.9,CN,001,1,Mermaid,2016],[24.38,CN,
        002,1,Monster Hunt,2015],[13.82,USA,003,2,Avatar,2010])
```

6. show

1）定义。

```
def show( ):Unit
def show( numRows:Int):Unit
```

2）功能描述。
- 不带参数时，用表格的形式显示 DataFrame 的前 20 行记录。
- 指定参数 numRows 时，用表格的形式显示 DataFrame 指定的行数记录。

3）示例。

```
scala > df.show(3)
+--------+-------+---+----------+-------------+----+
|boxoffice|country| id|languageid|         name|year|
+--------+-------+---+----------+-------------+----+
|    33.9|     CN|001|         1|      Mermaid|2016|
|   24.38|     CN|002|         1| Monster Hunt|2015|
|   13.82|    USA|003|         2|       Avatar|2010|
+--------+-------+---+----------+-------------+----+
only showing top 3 rows
```

7. take

1）定义。

```
def take(n:Int):Array[Row]
```

2）功能描述。

类似 head 方法，返回 DataFrame 中指定的前 n 行的值。

3）示例。

```
scala > df.take(3)
res26:Array[org.apache.spark.sql.Row] = Array([33.9,CN,001,1,Mermaid,2016],[24.38,CN,
    002,1,Monster Hunt,2015],[13.82,USA,003,2,Avatar,2010])

scala > df.takeAsList(4)
16/04/27 14:37:37 INFOmapred.FileInputFormat:Total input paths to process :1
res27:java.util.List[org.apache.spark.sql.Row] = [[33.9,CN,001,1,Mermaid,2016],[24.38,
    CN,002,1,Monster Hunt,2015],[13.82,USA,003,2,Avatar,2010],[15.3,USA,004,2,
    Zootropolis,2016]]
```

8. cache

1）定义。

```
def cache():DataFrame.this.type
```

2）功能描述。

将 DataFrame 缓存到内存中。

3）示例。

```
scala > df.cache
```

```
res28: org.apache.spark.sql.DataFrame = [boxoffice: double, country: string, id: string, languageid:
         bigint, name: string, year: bigint]
```

9. columns

1)定义。

```
def columns: Array[String]
```

2)功能描述。

以数组形式返回 DataFrame 的所有列名。

3)示例。

```
scala > df.columns
res29: Array[String] = Array(boxoffice, country, id, languageid, name, year)
```

10. dtypes

1)定义。

```
def dtypes: Array[(String, Stirng)]
```

2)功能描述。

以数组形式返回所有列名及其对应数据类型。

3)示例。

```
scala > df.dtypes
res30: Array[(String, String)] = Array((boxoffice, DoubleType), (country, StringType), (id, String-
         Type), (languageid, LongType), (name, StringType), (year, LongType))
```

11. explain

1)定义。

```
def explain(extended: Boolean): Unit
```

2)功能描述。

这个方法用于调试目的。
- 不带参数时,仅将 DataFrame 的物理计划打印到 Web 控制台上。
- 当指定参数 extended 为 true 时,打印所有计划到 Web 控制台上,包括解析逻辑计划、分析逻辑计划、优化的逻辑计划和物理计划。

3)示例。
- 不带参数时,仅将物理计划打印到 Web 控制台上。

```
scala > df.explain
== Physical Plan ==
```

第2章 DataFrame原理与常用操作

```
InMemoryColumnarTableScan [boxoffice#0,country#1,id#2,languageid#3L,name#4,year#5L],InMem-
    oryRelation [boxoffice#0,country#1,id#2,languageid#3L,name#4,year#5L],true,10000,Stor-
    ageLevel(true,true,false,true,1),Scan JSONRelation[boxoffice#0,country#1,id#2,languageid#
    3L,name#4,year#5L] InputPaths:hdfs://Master:9000/library/SparkSQL/Data/film.json,None
```

- 指定参数 extended 为 true 时，打印所有计划到 Web 控制台上。

```
scala > df.explain(true)
== Parsed Logical Plan ==
Relation[boxoffice#0,country#1,id#2,languageid#3L,name#4,year#5L] JSONRelation

== Analyzed Logical Plan ==
boxoffice:double,country:string,id:string,languageid:bigint,name:string,year:bigint
Relation[boxoffice#0,country#1,id#2,languageid#3L,name#4,year#5L] JSONRelation

== Optimized Logical Plan ==
InMemoryRelation [boxoffice#0,country#1,id#2,languageid#3L,name#4,year#5L],true,10000,Stor-
    ageLevel(true,true,false,true,1),Scan JSONRelation[boxoffice#0,country#1,id#2,languageid#
    3L,name#4,year#5L] InputPaths:hdfs://Master:9000/library/SparkSQL/Data/film.json,None

== Physical Plan ==
InMemoryColumnarTableScan [boxoffice#0,country#1,id#2,languageid#3L,name#4,year#5L],InMem-
    oryRelation [boxoffice#0,country#1,id#2,languageid#3L,name#4,year#5L],true,10000,Stor-
    ageLevel(true,true,false,true,1),Scan JSONRelation[boxoffice#0,country#1,id#2,languageid#
    3L,name#4,year#5L] InputPaths:hdfs://Master:9000/library/SparkSQL/Data/film.json,None
```

12. printSchema

1）定义。

```
def printSchema():Unit
```

2）功能描述。

以树形结构将 DataFrame 的 Schema 信息打印到 Web 控制台上。

3）示例。

```
scala > df.printSchema
root
 |-- boxoffice:double (nullable = true)
 |-- country:string (nullable = true)
 |-- id:string (nullable = true)
 |-- languageid:long (nullable = true)
 |-- name:string (nullable = true)
 |-- year:long (nullable = true)
```

13. registerTempTable

1）定义。

```
def registerTempTable(tableName:String):Unit
```

2）功能描述。

将 DataFrame 注册为指定名字的临时表。

3）示例。

```
scala > df.registerTempTable("film")

////注册成临时表之后,可以使用 SQLContext 的 sql 方法查询该临时表
scala > val oldfilm = sqlContext.sql("SELECT name FROM film WHERE year < 2013")
oldfilm:org.apache.spark.sql.DataFrame = [name:string]

scala > oldfilm.show
16/04/27 14:43:10 INFOmapred.FileInputFormat:Total input paths to process :1
+----------+
|   name   |
+----------+
|  Avatar  |
|First Love|
+----------+
```

4）解析。

将 DataFrame 注册成临时表之后，可以使用 SQLContext 的 sql 方法，对其执行 SQL 语句。

14. schema

1）定义。

```
def schema:StructType
```

2）功能描述。

返回 DataFrame 的 Schema 信息，对应类型为 StrutType。

3）示例。

```
scala > df.schema
res36:org.apache.spark.sql.types.StructType = StructType(StructField(boxoffice,DoubleType,true),
    StructField(country,StringType,true),StructField(id,StringType,true),StructField(languageid,
    LongType,true),StructField(name,StringType,true),StructField(year,LongType,true))
```

15. toDF

1）定义。

```
def toDF():DataFrame
```

第2章 DataFrame原理与常用操作

```
def toDF(colNames:String * ):DataFrame
```

2）功能描述。
- 不带参数的 toDF 返回它本身。
- 带字符串数组的参数时，返回新的 DataFrame，该 DataFrame 重命名了各列名。

3）示例。

```
scala > val newToDf = df.toDF
newToDf:org.apache.spark.sql.DataFrame = [boxoffice:double,country:string,id:string,languageid:
    bigint,name:string,year:bigint]

scala > val newToDf2 = df.toDF("t1","t2","t3","test1","test2","test3")
newToDf2:org.apache.spark.sql.DataFrame = [t1:double, t2:string, t3:string, test1:bigint, test2:
    string,test3:bigint]
```

16. persist

1）定义。

```
def persist(newLevel:StorageLevel):DataFrame.this.type
def persist():DataFrame.this.type
def unpersist():DataFrame.this.type
def unpersist(blocking:Boolean):DataFrame.this.type
```

2）功能描述。

以给定的存储等级将 DataFrame 持久化到内存或者磁盘中。unpersist 则是将 DataFrame 标记为非持久化的。
- persist(newLevel:StorageLevel)：设置 RDD 的存储级别，在其首次进行计算以后持久化值。如 RDD 没设置存储级别，此方法用于设置新的存储级别。本地检查点将会提示异常。
- persist()：以默认的存储级别（MEMORY_ONLY）持久化 RDD。
- unpersist()：设置 RDD 为非持久化，其中 unpersist 的入参 blocking 默认设置为 true，即阻塞直到所有块被删除。
- unpersist(blocking:Boolean)：设置 RDD 为非持久化，清除 RDD 在内存和磁盘中的所有块。

3）示例。

```
scala > df.persist(org.apache.spark.storage.StorageLevel.MEMORY_ONLY)
16/04/27 14:45:46 WARN execution.CacheManager:Asked to cache already cached data.
res37:org.apache.spark.sql.DataFrame = [boxoffice:double, country:string, id:string, languageid:
    bigint,name:string,year:bigint]

scala > df.unpersist
```

```
res38: org. apache. spark. sql. DataFrame = [boxoffice: double, country: string, id: string, languageid:
bigint, name: string, year: bigint]

scala > df. unpersist( true)
res39: org. apache. spark. sql. DataFrame = [boxoffice: double, country: string, id: string, languageid:
    bigint, name: string, year: bigint]
```

17. agg

1）定义。

```
def agg( expr: Column, exprs: Column * ) : DataFrame
def agg( exprs: java. util. Map[ String, String ] ) : DataFrame
def agg( exprs: Map[ String, String ] ) : DataFrame
def agg( aggExpr: ( String, String ) , aggExprs: ( String, String ) * ) : DataFrame
```

2）功能描述。

agg 是 Spark1.5.x 开始提供的一类内置函数。agg 这一系列的方法，为 DataFrame 提供数据列不需要经过 group 就可以执行统计操作。

3）示例。

下面的示例是统计 year 的最大值和 boxoffice 的平均值。

```
////显示前面准备的电影票房收入数据
scala > film. show
+---------+--------+---+----------+-----------+----+
| boxoffice| country| id| languageid|       name|year|
+---------+--------+---+----------+-----------+----+
|     33.9|      CN|001|         1|    Mermaid|2016|
|    24.38|      CN|002|         1|Monster Hunt|2015|
|    13.82|     USA|003|         2|     Avatar|2010|
|     15.3|     USA|004|         1| Zootropolis|2016|
|     5.33|      Uk|005|         2|    Spectre|2015|
|     1.54|      UK|006|         2|   Sherlock|2016|
|     1.54|      TH|007|         3| First Love|2012|
+---------+--------+---+----------+-----------+----+

////统计最大年份、平均票房收入
scala > df. agg( max($ "year" ) , avg($ "boxoffice" ) )
res3: org. apache. spark. sql. DataFrame = [max( year) : bigint, avg( boxoffice) : double]

scala > df. agg( max($ "year" ) , avg($ "boxoffice" ) ). show
+---------+--------------+
| max( year)| avg( boxoffice)|
```

```
+----------+------------------+---+
|   2016   |13.687142857142856| g |
+----------+------------------+---+
```

////直接使用 agg 方法和先用 groupBy 分组再调用 agg 方法的结果是一样的
scala > df.groupBy().agg(max($"year"),avg($"boxoffice")).show
```
+---------+------------------+
|max(year)|   avg(boxoffice) |
+---------+------------------+
|   2016  |13.687142857142856|
+---------+------------------+
```

下面是使用 Map 作为参数的示例，分别统计 year 的最小值和 boxoffice 的平均值。

scala > df.agg(Map("year" -> "min","boxoffice" -> "mean"))
res6:org.apache.spark.sql.DataFrame = [min(year):bigint,avg(boxoffice):double]

scala > df.agg(Map("year" -> "min","boxoffice" -> "mean")).show
```
+---------+------------------+
|min(year)|   avg(boxoffice) |
+---------+------------------+
|   2010  |13.687142857142856|
+---------+------------------+
```

上面的代码与下面使用二元数组作为参数的示例相同。

scala > df.agg(("year" -> "min"),("boxoffice" -> "mean"))
res8:org.apache.spark.sql.DataFrame = [min(year):bigint,avg(boxoffice):double]

scala > df.agg(("year" -> "min"),("boxoffice" -> "mean")).show
```
+---------+------------------+
|min(year)|   avg(boxoffice) |
+---------+------------------+
|   2010  |13.687142857142856|
+---------+------------------+
```

18. apply

1）定义。

```
def apply(colName:String):Column
```

2）功能描述。

根据指定列名返回 DataFrame 的列，其类型为 Column。

3）示例。

```
scala > df("year")
res10:org.apache.spark.sql.Column = year

scala > df.col("year")
res12:org.apache.spark.sql.Column = year
```

19. as

1）定义。

```
def as(alias:Symbol):DataFrame
```

2）功能描述。

调用 as 方法后，使用别名构建 DataFrame。

3）示例。

首先，注册临时表，然后修改调试日志的级别，方便查看调试信息。

```
////注册临时表
scala > film.registerTempTable("film")

////修改调试日志的级别
scala > Logger.getLogger("org.apache.spark.sql").setLevel(Level.DEBUG)
```

为了分析这个方法的作用，下面分别查看带 as 方法和不带 as 方法的两种情况。
- 不带 as 方法时的调试信息：

```
scala > val sss = sqlContext.sql("select year,boxoffice from film").explain(true)
16/04/26 19:46:06 INFO parse.ParseDriver:Parsing command:select year,boxoffice from film
16/04/26 19:46:07 INFO parse.ParseDriver:Parse Completed
16/04/26 19:46:07 DEBUG analysis.Analyzer $ ResolveReferences:Resolving 'year to year#5L
16/04/26 19:46:07 DEBUG analysis.Analyzer $ ResolveReferences:Resolving 'boxoffice to boxoffice#0
16/04/26 19:46:07 DEBUG hive.HiveContext $ $ anon $ 3:
=== Result of Batch Resolution ===
! 'Project [unresolvedalias('year),unresolvedalias('boxoffice)]    Project [year#5L,boxoffice#0]
!  +- 'UnresolvedRelation 'film, None                              +- Subquery film
!                                                                     +- Relation
[boxoffice#0,country#1,id#2,languageid#3L,name#4,year#5L] JSONRelation

16/04/26 19:46:07 DEBUG analysis.Analyzer $ ResolveReferences:Resolving 'year to year#5L
16/04/26 19:46:07 DEBUG analysis.Analyzer $ ResolveReferences:Resolving 'boxoffice to boxoffice#0
16/04/26 19:46:07 DEBUG hive.HiveContext $ $ anon $ 3:
=== Result of Batch Resolution ===
```

第2章 DataFrame原理与常用操作

```
!       'Project [unresolvedalias('year),unresolvedalias('boxoffice)]        Project [year#5L,boxoffice#0]
!       +- 'UnresolvedRelation 'film',None                                   +- Subquery film
!                                                                               +- Relation
[boxoffice#0,country#1,id#2,languageid#3L,name#4,year#5L] JSONRelation

16/04/26 19:46:07 DEBUG optimizer.DefaultOptimizer:
=== Result of Batch RemoveSubQueries ===
Project [year#5L,boxoffice#0]                                                Project [year#5L,boxoffice#0]
!       +- Subquery film
            +- Relation[boxoffice#0,country#1,id#2,languageid#3L,name#4,year#5L] JSONRelation
!       +- Relation[boxoffice#0,country#1,id#2,languageid#3L,name#4,year#5L] JSONRelation


== Parsed Logical Plan ==
'Project [unresolvedalias('year),unresolvedalias('boxoffice)]
+- 'UnresolvedRelation 'film',None

== Analyzed Logical Plan ==
year:bigint,boxoffice:double
Project [year#5L,boxoffice#0]
+- Subquery film
    +- Relation[boxoffice#0,country#1,id#2,languageid#3L,name#4,year#5L] JSONRelation

== Optimized Logical Plan ==
Project [year#5L,boxoffice#0]
+- Relation[boxoffice#0,country#1,id#2,languageid#3L,name#4,year#5L] JSONRelation

== Physical Plan ==
ScanJSONRelation[year#5L,boxoffice#0] InputPaths:hdfs://Master:9000/library/SparkSQL/Data/film.json
sss:Unit = ()
```

- 带as方法时的调试信息:

```
scala > val sss = sqlContext.sql("select year,boxoffice from film").as("alise").explain(true)
16/04/26 19:47:21 INFO parse.ParseDriver:Parsing command:select year,boxoffice from film
16/04/26 19:47:21 INFO parse.ParseDriver:Parse Completed
16/04/26 19:47:21 DEBUG analysis.Analyzer$ResolveReferences:Resolving 'year to year#5L
16/04/26 19:47:21 DEBUG analysis.Analyzer$ResolveReferences:Resolving 'boxoffice to boxoffice#0
16/04/26 19:47:21 DEBUG hive.HiveContext$$anon$3:
=== Result of Batch Resolution ===
!       'Project [unresolvedalias('year),unresolvedalias('boxoffice)]        Project [year#5L,boxoffice#0]
!       +- 'UnresolvedRelation 'film',None                                   +- Subquery film
```

```
              !                                                        +- Relation
   [boxoffice#0,country#1,id#2,languageid#3L,name#4,year#5L] JSONRelation

   16/04/26 19:47:21 DEBUG optimizer.DefaultOptimizer:
   === Result of Batch RemoveSubQueries ===
   ! Subquery alise
           Project [year#5L,boxoffice#0]
   ! +- Project [year#5L,boxoffice#0]
           +- Relation[boxoffice#0,country#1,id#2,languageid#3L,name#4,year#5L] JSONRelation
   !       +- Subquery film

   !         +- Relation[boxoffice#0,country#1,id#2,languageid#3L,name#4,year#5L] JSONRelation

   == Parsed Logical Plan ==
   Subquery alise
   +- Project [year#5L,boxoffice#0]
      +- Subquery film
         +- Relation[boxoffice#0,country#1,id#2,languageid#3L,name#4,year#5L] JSONRelation

   == Analyzed Logical Plan ==
   year:bigint,boxoffice:double
   Subquery alise
   +- Project [year#5L,boxoffice#0]
      +- Subquery film
         +- Relation[boxoffice#0,country#1,id#2,languageid#3L,name#4,year#5L] JSONRelation

   == Optimized Logical Plan ==
   Project [year#5L,boxoffice#0]
   +- Relation[boxoffice#0,country#1,id#2,languageid#3L,name#4,year#5L] JSONRelation

   == Physical Plan ==
   ScanJSONRelation[year#5L,boxoffice#0] InputPaths:hdfs://Master:9000/library/SparkSQL/Data/
   film.json
   sss:Unit = ()
```

4）解析。

通过上面两种情况的比较可以看出，仅在解析逻辑计划、分析逻辑计划中，使用了别名 Subquery alise。

20. distinct

1）定义。

```
def distinct():DataFrame
```

2)功能描述。

返回对 DataFrame 的数据记录去重后的 DataFrame。

3)示例。

```
scala > df.select("year").distinct.show
+----+
|year|
+----+
|2010|
|2012|
|2015|
|2016|
+----+
```

4)解析。

在实例中,选择了有重复数据记录的"year"列,最后调用 distinct 方法进行去重。

21. except

1)定义。

```
def except(other:DataFrame):DataFrame
```

2)功能描述。

返回 DataFrame,包含当前 DataFrame 的数据记录,同时这些 Rows 不在另一个 DataFrame 中,相当于两个 DataFrame 做减法。

3)示例。

```
////以表格形式查看全部电影的票房数据
scala > df.show
+--------+-------+---+---------+------------+----+
|boxoffice|country| id|languageid|        name|year|
+--------+-------+---+---------+------------+----+
|    33.9|     CN|001|        1|     Mermaid|2016|
|   24.38|     CN|002|        1|Monster Hunt|2015|
|   13.82|    USA|003|        2|      Avatar|2010|
|    15.3|    USA|004|        2|  Zootropolis|2016|
|    5.33|     Uk|005|        2|     Spectre|2015|
|    3.82|     UK|006|        2|    Sherlock|2016|
|    1.54|     TH|007|        3|  First Love|2012|
+--------+-------+---+---------+------------+----+

////以表格形式查看最新电影的票房数据
```

```
scala > newDf.show
16/04/27 09:11:51 INFOmapred.FileInputFormat:Total input paths to process :1
+---------+-------+----+----------+--------------+----+
| boxoffice | country | id | languageid |     name     | year |
+---------+-------+----+----------+--------------+----+
|   33.9   |  CN   | 001 |    1     |   Mermaid    | 2016 |
|   16.79  |  CN   | 008 |    1     |  The Ghouls  | 2015 |
|   13.82  |  CN   | 009 |    1     | Breakup Buddie | 2014 |
+---------+-------+----+----------+--------------+----+
```

////df 和 newDf 做排除运算

```
scala > df.except(newDf).show
+---------+-------+----+----------+--------------+----+
| boxoffice | country | id | languageid |     name     | year |
+---------+-------+----+----------+--------------+----+
|  24.38   |  CN   | 002 |    1     | Monster Hunt | 2015 |
|  13.82   |  USA  | 003 |    2     |    Avatar    | 2010 |
|  15.3    |  USA  | 004 |    2     |  Zootropolis | 2016 |
|   5.33   |  Uk   | 005 |    2     |   Spectre    | 2015 |
|   3.82   |  UK   | 006 |    2     |   Sherlock   | 2016 |
|   1.54   |  TH   | 007 |    3     |  First Love  | 2012 |
+---------+-------+----+----------+--------------+----+
```

4）解析。

因为 001Mermaid 这部电影既在 df 中，又在 newDf 中，所以经过排除运算之后，001 这条记录不再被显示。

22. filter

1）定义。

```
def filter(conditionExpr:String):DataFrame
def filter(condition:Column):DataFrame
```

2）功能描述。

按参数指定的 SQL 表达式的条件过滤 DataFrame。
- filter(conditionExpr:String)根据给的的 SQL 表达式进行过滤；filter(condition:Column)根据给的条件进行过滤。例如过滤出年龄大于 15 岁的用户记录：

```
peopleDf.filter("age>15")
peopleDs.where($"age">15)
```

这两种写法是等价的。

第2章 DataFrame原理与常用操作

3）示例。

```
////过滤大于2015年的记录
scala > df.filter("year > 2015").show
+--------+-------+---+----------+----------+----+
|boxoffice|country| id|languageid|      name|year|
+--------+-------+---+----------+----------+----+
|    33.9|     CN|001|         1|   Mermaid|2016|
|    15.3|    USA|004|         2|Zootropolis|2016|
|    3.82|     UK|006|         2|   Sherlock|2016|
+--------+-------+---+----------+----------+----+

////过滤大于2015年,并且国家为中国的记录
scala > df.where($"year" > 2015 && $"country" === "CN").show
+--------+-------+---+----------+-------+----+
|boxoffice|country| id|languageid|   name|year|
+--------+-------+---+----------+-------+----+
|    33.9|     CN|001|         1|Mermaid|2016|
+--------+-------+---+----------+-------+----+
```

23. groupBy

1）定义。

```
def groupBy(col1:String,cols:String*):GroupedData
def groupBy(cols:Column*):GroupedData
```

2）功能描述。

使用一个或多个指定的列对DataFrame进行分组,以便对它们执行聚合操作。

3）示例。

```
////根据country列对df进行分组,求年份的最大值和票房的平均值
scala > df.groupBy("country").agg(
      | "year" -> "max",
      | "boxoffice" -> "mean"
      | ).show
+-------+---------+--------------+
|country|max(year)|avg(boxoffice)|
+-------+---------+--------------+
|     TH|     2012|          1.54|
|     UK|     2016|          3.82|
|     Uk|     2015|          5.33|
```

```
| USA | 2016 | 14.56 |
| CN  | 2016 | 29.14 |
+-------+---------+---------------+
```

示例中先根据 country 列对 df 进行分组，分组后求年份的最大值和票房的平均值。

24. intersect

1）定义。

```
def intersect(other:DataFrame):DataFrame
```

2）功能描述。

取两个 DataFrame 中同时存在的数据记录，返回 DataFrame。

3）示例。

```
scala > df.select("name").intersect(newDf.select($"name")).show
+-------+
|  name |
+-------+
|Mermaid|
+-------+
```

在该示例中，因为 001 Mermaid 这部电影既在 df 中，又在 newDf 中，所以 intersect 的运算结果为 Mermaid。

25. join

1）定义。

```
def join(right:DataFrame,joinExprs:Column,joinType:String):DataFrame
def join(right:DataFrame,joinExprs:Column):DataFrame
def join(right:DataFrame,usingColumns:Seq[String],joinType:String):DataFrame
def join(right:DataFrame,usingColumns:Seq[String]):DataFrame
def join(right:DataFrame,usingColumn:String):DataFrame
def join(right:DataFrame):DataFrame
```

2）功能描述。

对两个 DataFrame 执行 join 操作。join 根据传入参数的不同有多种实现方法。不带参数时取笛卡儿积，仅带 join Exprs 时默认为 Inner Join，第三个 join 参数 joinType 可以指定具体的 join 操作，例如：'inner'、'left outer'、'rightouter'、'leftsemio'。

3）示例。

在本章的 2.2.3 节的第 14 示例中，已经介绍了如何对两个 DataFrame 实例进行 join 操作。下面再来做一个简单的 join 示例，将影片信息和语言信息做 join 关联操作，查询显示汉语、英语其他语种的电影影片信息。

第2章 DataFrame原理与常用操作

```
////查看语言信息数据
scala > language.show
+---------+-----------+
| language | languageid |
+---------+-----------+
| chinese | 1 |
| english | 2 |
|   other | 3 |
+---------+-----------+

////将影片信息和语言信息做外联操作
scala > film.join(language,film("languageid") === language("languageid"),"outer").show
+---------+--------+-----+-----------+-------------+------+---------+-----------+
| boxoffice | country | id | languageid |       name | year | language | languageid |
+---------+--------+-----+-----------+-------------+------+---------+-----------+
|  33.9 | CN | 001 | 1 |      Mermaid | 2016 | chinese | 1 |
| 24.38 | CN | 002 | 1 | Monster Hunt | 2015 | chinese | 1 |
|  5.33 | Uk | 005 | 2 |      Spectre | 2015 | english | 2 |
|  3.82 | UK | 006 | 2 |     Sherlock | 2016 | english | 2 |
| 13.82 | USA | 003 | 2 |       Avatar | 2010 | english | 2 |
|  15.3 | USA | 004 | 2 |   Zootropolis | 2016 | english | 2 |
|  1.54 | TH | 007 | 3 |   First Love | 2012 |   other | 3 |
+---------+--------+-----+-----------+-------------+------+---------+-----------+
```

26. limit

1）定义。

```
def limit(n:Int):DataFrame
```

2）功能描述。

返回 DataFrame 的前 n 条数据记录。

3）示例

```
scala > df.limit(3).show
+---------+--------+-----+-----------+-------------+------+
| boxoffice | country | id | languageid |       name | year |
+---------+--------+-----+-----------+-------------+------+
|  33.9 | CN | 001 | 1 |      Mermaid | 2016 |
| 24.38 | CN | 002 | 1 | Monster Hunt | 2015 |
| 13.82 | USA | 003 | 2 |       Avatar | 2010 |
+---------+--------+-----+-----------+-------------+------+
```

27. orderBy 和 sort

1）定义。

```
def sort(sortExprs:Column *):DataFrame
def sort(sortCol:String,sortCols:String *):DataFrame
def orderBy(sortExprs:Column *):DataFrame
def orderBy(sortCol:String,sortCols:String *):DataFrame
```

2）功能描述。

对 DataFrame 按指定的一列或多列进行排序，分别支持字符串或 Column 的参数列表。

- sort(sortExprs:Column *)：根据给定的表达式返回一个新的 DataFrame。例如：

```
df.sort($"col1",$"col2".desc)
```

- sort(sortCol:String,sortCols:String *)：根据指定的列返回一个新的 DataFrame，所有列升序排列。以下三种写法等价：

```
df.sort("sortcol")
df.sort($"sortcol")
df.sort($"sortcol".asc)
```

- orderBy(sortExprs:Column *)：按给定的表达式返回一个新的 DataFrame。这是 sort 排序函数的别名。输入参数为多个 Column 类。
- orderBy(sortCol:String,sortCols:String *)：按给定的表达式返回一个新的 DataFrame。这是 sort 排序函数的别名。输入参数为多个 String 字符串。

3）示例。

```
////按 id 列顺序排序,并返回前 3 条记录
scala > df.sort("id").show(3)
+--------+-------+---+----------+-------------+----+
|boxoffice|country| id|languageid|         name|year|
+--------+-------+---+----------+-------------+----+
|    33.9 |    CN |001|         1|      Mermaid|2016|
|   24.38 |    CN |002|         1| Monster Hunt|2015|
|   13.82 |   USA |003|         2|       Avatar|2010|
+--------+-------+---+----------+-------------+----+
only showing top 3 rows

////按 id 列倒序排序,并返回前 3 条记录
scala > df.sort($"id".desc).show(3)
+--------+-------+---+----------+-------------+----+
|boxoffice|country| id|languageid|         name|year|
+--------+-------+---+----------+-------------+----+
|    1.54 |    TH |007|         3|   First Love|2012|
```

```
| 3.82 | UK  | 006 | 2 |      Sherlock | 2016 |
| 5.33 | Uk  | 005 | 2 |       Spectre | 2015 |
+------+-----+-----+---+---------------+------+
only showing top 3 rows
```

////按 year 和 boxoffice 两列顺序排序,即查询结果先按年份从小到大排序,其中 2016 年有 3 条记录,再将此三条记录按照票房收入从低到高排列,最终返回排序以后的记录

scala > df.orderBy("year","boxoffice").show
16/04/27 09:50:41 INFOmapred.FileInputFormat:Total input paths to process :1

```
+---------+--------+----+-----------+-------------+------+
| boxoffice| country| id | languageid|         name| year |
+---------+--------+----+-----------+-------------+------+
|    13.82 |    USA | 003|          2|       Avatar| 2010 |
|     1.54 |     TH | 007|          3|  First Love | 2012 |
|     5.33 |     Uk | 005|          2|      Spectre| 2015 |
|    24.38 |     CN | 002|          1| Monster Hunt| 2015 |
|     3.82 |     UK | 006|          2|     Sherlock| 2016 |
|    15.3  |    USA | 004|          2|   Zootropolis| 2016 |
|    33.9  |     CN | 001|          1|      Mermaid| 2016 |
+---------+--------+----+-----------+-------------+------+
```

////按 year 和 boxoffice 两列排序的另外一种写法,col("year") 及 df("boxoffice") 是另外一种的语法表达方式

scala > df.orderBy(col("year"),df("boxoffice")).show
16/04/27 09:51:28 INFOmapred.FileInputFormat:Total input paths to process :1

```
+---------+--------+----+-----------+-------------+------+
| boxoffice| country| id | languageid|         name| year |
+---------+--------+----+-----------+-------------+------+
|    13.82 |    USA | 003|          2|       Avatar| 2010 |
|     1.54 |     TH | 007|          3|  First Love | 2012 |
|     5.33 |     Uk | 005|          2|      Spectre| 2015 |
|    24.38 |     CN | 002|          1| Monster Hunt| 2015 |
|     3.82 |     UK | 006|          2|     Sherlock| 2016 |
|    15.3  |    USA | 004|          2|   Zootropolis| 2016 |
|    33.9  |     CN | 001|          1|      Mermaid| 2016 |
+---------+--------+----+-----------+-------------+------+
```

28. sample（取样）

1）定义。

```
def sample(withReplacement:Boolean,fraction:Double):DataFrame
def sample(withReplacement:Boolean,fraction:Double,seed:Long):DataFrame
```

2) 功能描述。

Sample 对 RDD 中的数据集进行采样，生成一个新的 RDD。withReplacement = true，表示重复抽样；withReplacement = false，表示不重复抽样；fraction 参数是生成行的比例。

- sample（withReplacement：Boolean，fraction：Double）：使用随机因子对 DataFrame 的 Rows 进行取样，返回一个新的 DataFrame。
- sample（withReplacement：Boolean，fraction：Double，seed：Long）：按指定因子（seed）对 DataFrame 的 Rows 进行取样，返回一个新的 DataFrame。

3) 示例。

```
scala > df.sample(false,0.5).show
16/04/27 10:17:21 INFOmapred.FileInputFormat:Total input paths to process :1
+--------+-------+---+----------+-----------+----+
|boxoffice|country| id|languageid|       name|year|
+--------+-------+---+----------+-----------+----+
|   24.38 |     CN|002|         1|Monster Hunt|2015|
|   13.82 |    USA|003|         2|      Avatar|2010|
|   15.3  |    USA|004|         2|  Zootropolis|2016|
|    5.33 |     Uk|005|         2|     Spectre|2015|
|    1.54 |     TH|007|         3|  First Love|2012|
+--------+-------+---+----------+-----------+----+


scala > df.sample(false,0.5,1).show
+--------+-------+---+----------+-----------+----+
|boxoffice|country| id|languageid|       name|year|
+--------+-------+---+----------+-----------+----+
|   33.9  |     CN|001|         1|     Mermaid|2016|
|   13.82 |    USA|003|         2|      Avatar|2010|
|   15.3  |    USA|004|         2|  Zootropolis|2016|
|    5.33 |     Uk|005|         2|     Spectre|2015|
|    3.82 |     UK|006|         2|    Sherlock|2016|
|    1.54 |     TH|007|         3|  First Love|2012|
+--------+-------+---+----------+-----------+----+


scala > df.sample(true,0.5,1).show
+--------+-------+---+----------+-----------+----+
|boxoffice|country| id|languageid|       name|year|
+--------+-------+---+----------+-----------+----+
|   33.9  |     CN|001|         1|     Mermaid|2016|
|   13.82 |    USA|003|         2|      Avatar|2010|
|   15.3  |    USA|004|         2|  Zootropolis|2016|
+--------+-------+---+----------+-----------+----+
```

第2章 DataFrame原理与常用操作

当withReplacement为true时,采用PossionSampler抽样器(Possion,泊松分布);当withReplacemet为false时,采用BernoulliSampler抽样器(Bernoulli,伯努利采样)

29. select

////写法一:选取name、boxoffice两列,输出数据
```
scala > df.select("name","boxoffice").show
+------------+---------+
|        name|boxoffice|
+------------+---------+
|     Mermaid|     33.9|
|Monster Hunt|    24.38|
|      Avatar|    13.82|
|  Zootropolis|    15.3|
|     Spectre|     5.33|
|    Sherlock|     3.82|
|  First Love|     1.54|
+------------+---------+
```

////写法二:选取name、country两列,输出数据
```
scala > df.select($"name",$"country").show
+------------+-------+
|        name|country|
+------------+-------+
|     Mermaid|     CN|
|Monster Hunt|     CN|
|      Avatar|    USA|
|  Zootropolis|   USA|
|     Spectre|     Uk|
|    Sherlock|     UK|
|  First Love|     TH|
+------------+-------+
```

////写法三:选取name、country两列,输出数据
```
scala > df.select(col("name"),df("country")).show
+------------+-------+
|        name|country|
+------------+-------+
|     Mermaid|     CN|
|Monster Hunt|     CN|
|      Avatar|    USA|
|  Zootropolis|   USA|
|     Spectre|     Uk|
|    Sherlock|     UK|
|  First Love|     TH|
```

```
+------------+-------+
```

////写法四:以计算表达式方式选取列,然后输出数据。本示例选取下面3列:
////year + 10、name 列改成 filmName、languageid 取绝对值
scala > df.selectExpr("year + 10","name as filmName","abs(languageid)").show
```
+----------+------------+---------------+
|(year + 10)| filmName  |abs(languageid)|
+----------+------------+---------------+
|   2026   |   Mermaid  |       1       |
|   2025   | Monster Hunt|      1       |
|   2020   |   Avatar   |       2       |
|   2026   | Zootropolis|       2       |
|   2025   |   Spectre  |       2       |
|   2026   |  Sherlock  |       2       |
|   2022   | First Love |       3       |
+----------+------------+---------------+
```

30. unionAll

1)定义。

```
def unionAll(other:DataFrame):DataFrame
```

2)功能描述。

合并两个 DataFrame 的全部数据记录。

3)示例。

```
////合并 df 和 newDf 的全部记录
scala > df.unionAll(newDf).show
+--------+-------+---+----------+------------+----+
|boxoffice|country| id|languageid|    name    |year|
+--------+-------+---+----------+------------+----+
|  33.9  |   CN  |001|    1     |   Mermaid  |2016|
|  24.38 |   CN  |002|    1     |Monster Hunt|2015|
|  13.82 |  USA  |003|    2     |   Avatar   |2010|
|  15.3  |  USA  |004|    2     | Zootropolis|2016|
|  5.33  |   Uk  |005|    2     |   Spectre  |2015|
|  3.82  |   UK  |006|    2     |  Sherlock  |2016|
|  1.54  |   TH  |007|    3     | First Love |2012|
|  33.9  |   CN  |001|    1     |   Mermaid  |2016|
|  16.79 |   CN  |008|    1     | The Ghouls |2015|
```

```
|13.82   |CN     |009|1         |Breakup Buddie|2014|
+--------+-------+---+----------+--------------+----+
```

31. withColumn 和 withColumnRenamed

1）定义。

```
def withColumn(colName:String,col:Column):DataFrame
def withColumnRenamed(existingName:String,newName:String):DataFrame
```

2）功能描述。

对 DataFrame 列进行操作，withColumn 增加 DataFrame 的列信息，withColumnRenamed 则是对 DataFrame 的列进行重命名。

3）示例。

```
////新增 level 列,其值为 boxoffice 列/10
scala > df.withColumn("level",df("boxoffice")/10).show
+--------+-------+---+----------+-----------+----+------------------+
|boxoffice|country|id |languageid|name       |year|level             |
+--------+-------+---+----------+-----------+----+------------------+
|33.9    |CN     |001|1         |Mermaid    |2016|3.3899999999999997|
|24.38   |CN     |002|1         |Monster Hunt|2015|2.4379999999999997|
|13.82   |USA    |003|2         |Avatar     |2010|1.3820000000000001|
|15.3    |USA    |004|2         |Zootropolis|2016|1.53              |
|5.33    |Uk     |005|2         |Spectre    |2015|0.533             |
|3.82    |UK     |006|2         |Sherlock   |2016|0.382             |
|1.54    |TH     |007|3         |First Love |2012|0.154             |
+--------+-------+---+----------+-----------+----+------------------+

////将 boxoffice 列重命名为 level
scala > val rndf = df.withColumnRenamed("boxoffice","level")
rndf: org.apache.spark.sql.DataFrame = [level: double , country: string, id: string, languageid: bigint, name: string, year: bigint]

scala > rndf.show
+-----+-------+---+----------+-----------+----+
|level|country|id |languageid|name       |year|
+-----+-------+---+----------+-----------+----+
|33.9 |CN     |001|1         |Mermaid    |2016|
|24.38|CN     |002|1         |Monster Hunt|2015|
```

```
| 13.82 | USA | 003 | 2 |    Avatar   | 2010 |
| 15.3  | USA | 004 | 2 | Zootropolis | 2016 |
| 5.33  | Uk  | 005 | 2 |   Spectre   | 2015 |
| 3.82  | UK  | 006 | 2 |   Sherlock  | 2016 |
| 1.54  | TH  | 007 | 3 | First Love  | 2012 |
+-------+-----+-----+---+-------------+------+
```

32. insertInto、insertIntoJDBC、createJDBCTable

1）定义。

```
def insertInto(tableName:String):Unit
def insertInto(tableName:String,overwrite:Boolean):Unit
def insertIntoJDBC(url:String,table:String,overwrite:Boolean):Unit

def createJDBCTable(url:String,table:String,allowExisting:Boolean):Unit
```

2）功能描述。

- insertInto（tableName：String）：从 RDD 插入行到指定的表。如果表已经存在，则抛出异常。

- insertInto（tableName：String，overwrite：Boolean）：从 RDD 插入行到指定的表，可选择是否覆盖现有数据。

- insertIntoJDBC（url：String，table：String，overwrite：Boolean）：根据 url（参数 url 用来指定数据库信息）保存 DataFrame 到 JDBC 数据库的表中。如果表已经存在且模式兼容，overwrite 覆盖设置为 true，则在执行插入之前将先删除掉（truncate）表中的数据。表必须已经存在于数据库中；数据库表的模式和 RDD 的模式兼容，从 RDD 插入行，可以通过简单的声明，使用 INSERT INTO table VALUES（?，?，…，?）插入值，操作将不会失败。

- createJDBCTable（url：String，table：String，allowExisting：Boolean）：根据 url（参数 url 用来指定数据库信息）保存 DataFrame 到 JDBC 数据库的表中，将运行新建表（CREATE TABLE）和插入（INSERT INTO）的语句。如果设置 allowExisting 为 true，将删掉数据库中给定名称的表；如果设置 allowExisting 为 false，将抛出表已经存在的异常。

📖 createJDBCTable 和 insertIntoJDBC 从 Spark 1.4.0 版本开始使用，将在 Spark2.0 中废弃，可以使用 Write.jdbc()方法。

3）示例。

```
////创建 testin 这个 DataFrame 对象,把 1,3 这行数据存入其中
scala > val testin = sqlContext.sql("select 1,3").toDF("first","second")
```

testin:org.apache.spark.sql.DataFrame = [first:int,second:int]

scala > testin.registerTempTable("test")

////表名
scala > val dbtable = "TEST_JDBC"
dbtable:String = TEST_JDBC

////用 createJDBCTable 方法在 MySQL 的 Hive 数据库中创建一张表 dbtable,插入 testin 中的数据
scala > testin.createJDBCTable("jdbc:mysql://Master:3306/hive? user = root&password = root",
dbtable,true)

////创建 idf 这个 DataFrame 对象,把 2,4 这行数据存入其中
scala > val idf = sqlContext("select 2,4").toDF("first","second")
idf:org.apache.spark.sql.DataFrame = [first:int,second:int]

////再把 idf 中的数据通过 JDBC 写入到 MySQL 的 Hive 数据库的 dbtable 表中
scala > idf.insertIntoJDBC("jdbc:mysql://Master:3306/hive? user = root&password = root",dbtable,
true)

////通过 jdbc 从 MySQL 中读取 dbtable 表中的数据,存入 jdbcDF 这个 DataFrame 对象中
scala > val jdbcDF = sqlContext.load("jdbc",Map(
    "url" -> "jdbc:mysql://Master:3306/hive? user = root&password = root",
    "dbtable" -> dbtable))

jdbcDF:org.apache.spark.sql.DataFrame = [first:int,second:int]

////显示 jdbcDF 中的内容
scala > jdbcDF.show
+-----+------+
|first|second|
+-----+------+
| 1 | 3 |
| 2 | 4 |
+-----+------+

4) 解析。

该示例首先创建两个 DataFrame 对象:testin、idf。然后,testin 对象调用 createJD-BCTable 方法,在 MySQL 中创建一张表 dbtable,把 testin 中的数据写入其中。接下来 idf 对象调用 insertIntoJDBC 方法,将 idf 的数据插入到刚才创建的表 dbtable 中。

33. flatMap

1) 定义。

```
def flatMap[R](f:(Row)⇒TraversableOnce[R])(implicit arg0:ClassTag[R]):RDD[R]
```

2）功能描述。

创建一个新的 RDD 对 DataFrame 中的所有最后记录进行处理，并且将处理结果的所有数据仅返回一个数组对象。

flatMap[R]其中的，R 的类型是 ClassTag，ClassTag[T]通过 runtimeClass 清除给定的类型 T，这在 Array 元素类型未知，编译实例化元素特别有用。ClassTag 是 scala.reflect.api.TypeTags#TypeTag 的特殊情况，在运行时根据给定的类型封装，而 TypeTag 包含所有静态类型信息。ClassTag 是由 top – level 类构建，对于运行时创建 Array，这些信息足够了，因此，不必知道所有参数类型。

3）示例。

```
scala > val sdf = sqlContext.sql("select 1,2")
sdf:org.apache.spark.sql.DataFrame = [_c0:int,_c1:int]

scala > sdf.flatMap(x => List(x(0),x(1)))
res5:org.apache.spark.rdd.RDD[Any] = MapPartitionsRDD[25] at flatMap at <console>:29

scala > sdf.flatMap(x => List(x(0),x(1))).collect
res6:Array[Any] = Array(1,2)
```

示例中将数据记录转化为由每一列组成的 List。

34. foreach

1）定义。

```
def foreach(f:(Row)⇒Unit):Unit
def foreachPartition(f:(Iterator[Row])⇒Unit):Unit
```

2）功能描述。

foreach 方法对 DataFrame 中的数据记录进行循环遍历处理。foreachPartition 方法则是对对应分区中的数据记录进行处理，即 Iterator[Row]，使用方法类似。

3）示例。

```
scala > val sdf = sqlContext.sql("select 1,2")
sdf:org.apache.spark.sql.DataFrame = [_c0:int,_c1:int]

////foreach 打印 sdf 中的数据
scala > sdf.collect.foreach(x => println(x))
[1,2]

////foreach 循环输出分割线
scala > sdf.collect.foreach(x => println("-----------------------------------
---------------"))
```

第2章 DataFrame原理与常用操作

35. map

1）定义。

```
def map[R](f:(Row)⇒R)(implicit arg0:ClassTag[R]):RDD[R]
```

2）功能描述。

map 方法将 DataFrame 的数据记录按指定的函数映射成一个新的 RDD 实例。

3）示例。

```
scala > sdf.map(x => "First = " + x(0) + ";Second = " + x(1))
res10:org.apache.spark.rdd.RDD[String] = MapPartitionsRDD[27] at map at <console>:29

scala > sdf.map(x => "First = " + x(0) + ";Second = " + x(1)).collect().foreach(println)
First = 1;Second = 2
```

36. repartition

1）定义。

```
def repartition(numPartitions:Int):DataFrame
def repartition(partitionExprs:Column*):DataFrame
def repartition(numPartitions:Int,partitionExprs:Column*):DataFrame
```

2）功能描述

返回一个 DataFrame，该 DataFrame 按指定 numPartitions 对原 DataFrame 进行重分区。

- repartition（numPartitions：Int）：返回一个新的 DataFrame，生成 numPartitions 个分区。
- repartition（partitionExprs：Column*）：返回一个新的 DataFrame，根据给定的分区表达式保存现有的分区数，由此产生的 DataFrame 是哈希分区，这和 SQL（Hive QL）中的 DISTRIBUTE BY 操作是相同的。
- repartition（numPartitions：Int, partitionExprs：Column*）：返回一个新的 DataFrame，根据给定的分区表达式生成 numPartitions 个分区，由此产生的 DataFrame 是哈希分区，这和 SQL（Hive QL）中的 DISTRIBUTE BY 操作是相同的。

3）示例。

```
scala > df.repartition(1).rdd.partitions.size
res12:Int = 1
```

37. toJSON

1）定义。

```
def toJSON:RDD[String]
```

2）功能描述。

把 Dataframe 的数据记录用包含 JSON 字符串的 RDD 形式返回。

3）示例。

```
scala > df. toJSON. collect          //Collect 方法返回一个数组,包含 RDD 中所有的元素
res15:Array[String] =
Array({"boxoffice":33.9,"country":"CN","id":"001","languageid":1,"name":"Mermaid","year":2016},{"boxoffice":24.38,"country":"CN","id":"002","languageid":1,"name":"Monster Hunt","year":2015},{"boxoffice":13.82,"country":"USA","id":"003","languageid":2,"name":"Avatar","year":2010},{"boxoffice":15.3,"country":"USA","id":"004","languageid":2,"name":"Zootropolis","year":2016},{"boxoffice":5.33,"country":"Uk","id":"005","languageid":2,"name":"Spectre","year":2015},{"boxoffice":3.82,"country":"UK","id":"006","languageid":2,"name":"Sherlock","year":2016},{"boxoffice":1.54,"country":"TH","id":"007","languageid":3,"name":"First Love","year":2012})
```

38. queryExecution

1）定义。

```
val queryExecution:QueryExecution
```

2）功能描述。

返回 DataFrame 的查询执行语句，包含逻辑计划（Logical Plan）和物理计划（Physical Plan）。返回 DataFrame 的查询执行语句，包含逻辑计划和物理计划。逻辑计划描述了 DataFrame 生成数据所需的逻辑计算，Spark 查询优化器将优化逻辑计划，生成一个并行分布式有效执行的物理计划，DataFrame 可以通过 queryExecution 方法来查询逻辑计划和物理计划。

3）示例。

```
scala > df. queryExecution
res16:org. apache. spark. sql. execution. QueryExecution =
== Parsed Logical Plan ==
Relation[boxoffice#0,country#1,id#2,languageid#3L,name#4,year#5L] JSONRelation

== Analyzed Logical Plan ==
boxoffice:double,country:string,id:string,languageid:bigint,name:string,year:bigint
Relation[boxoffice#0,country#1,id#2,languageid#3L,name#4,year#5L] JSONRelation

== Optimized Logical Plan ==
Relation[boxoffice#0,country#1,id#2,languageid#3L,name#4,year#5L] JSONRelation

== Physical Plan ==
```

ScanJSONRelation[boxoffice#0,country#1,id#2,languageid#3L,name#4,year#5L] InputPaths:hdfs://Master:9000/library/SparkSQL/Data/film.json

2.6 本章小结

通过本章的学习，读者应该对 DataFrame 编程模型有了一个基本的了解，明确了 DataFrame 和 RDD 的区别，以及两者之间的转换。本章通过一系列常用 DataFrameAPI 的示例解析，详解了 DataFrame 的具体用法，在实际操作 DataFrame 的过程中，可以充分利用 SQL 的简洁性和 DataFrame 的功能特性，进行高效的数据处理。

第 3 章　Spark SQL 操作多种数据源

在 Spark SQL 中，对很多种数据格式的读取和保存方式都很简单。本章将通过具体的操作案例，一一介绍 Spark SQL 是如何读取和处理多种数据源的，包括：Parquet、Hive、Json、HBase 等。

3.1　通用的加载/保存功能

Spark SQL 对于 DataFrame 提供了数据加载和保存的操作方法：
- 数据加载：在 SQLContext 中通过 parquetFile、jsonFile、load、jdbc 等方法分别加载 parquet、json、文本文件或数据库的数据来创建 DataFrame。
- 保存：在 DataFrame 中通过 saveAsParquetFile、saveAsTable、save 把 DataFrame 中的数据保存到 Parquet、json、文本文件或数据库中。

Spark SQL 中的数据加载和保存是最基本也是比较重要的操作，接下来就通过案例演示具体的使用。

3.1.1　Spark SQL 加载数据

Spark SQL 加载数据是通过 load 函数实现的，下面就通过具体案例动手实践 load 函数的用法。实验的前提是启动 HDFS、Spark 集群，并且以集群的方式运行 Spark-Shell 命令行。

1. 创建目录

首先，在 HDFS 文件系统中创建 examples 目录，操作如下：

```
####在 HDFS 文件系统中创建 examples 目录
[root@ Master resources]#hdfs dfs -mkdir /examples

####显示 HDFS 下的目录
root@ master:~#hdfs dfs -ls /
Found 1 items
drwxr-xr-x   - root supergroup          0 2016-11-14 20:23 /examples
drwxr-xr-x   - root supergroup          0 2016-11-14 20:23 /data
```

2. 目录查询与确认

第一步的目录创建完成之后，可以通过 HDFS 的 Web 控制台界面，来查询新建目录是否已经创建成功。

通过在 Web 控制台输入 URL：http://Master:50070/explorer.html#/，或者直接输入 Master 的 IP 地址，如：http://xxx.xxx.xxx.xxx:50070/explorer.html#/，即可进入 HDFS 的 Web 控制台界面，如图 3-1 所示。由此可以确认上一步中新建的目录 examples 已经创建成功。

图 3-1　HDFS Web 控制台界面

3. 文件传入

确认目录创建好之后，将 Spark 系统中自带的 examples\src\main\resource 中的 people.json 文件上传到 HDFS 的 examples 目录下。具体操作如下：

```
[root@ Master resources]#hdfs dfs – put people.json /examples
```

4. 文件读取

将文件传至新目录 examples 下后，可以通过以下操作读取 HDFS 目录下的 examples 中的 people.json 文件。

```
scala > sqlContext
res0:org.apache.spark.sql.SQLContext = org.apache.spark.sql.hive.HiveContext@274b0c2a

////用 format 指定读取文件的格式
scala > val peopleDF = sqlContext.read.format("json").load("/examples/people.json")
16/04/30 16:41:22 INFOjson.JSONRelation: Listing hdfs://Master:9000/examples/people.json on driver
peopleDF:org.apache.spark.sql.DataFrame = [age:bigint, name:string]
```

5. 数据显示

读取文件之后，可以使用 show 命令将文件中的数据显示出来。具体实现如下：

```
scala > peopleDF.show
+----+-------+
| age|   name|
+----+-------+
|null|Michael|
|  30|   Andy|
```

```
|  19 | Justin |
+----+--------+
```

3.1.2 Spark SQL 保存数据

Spark SQL 通过 save 函数可以将操作完的结果指定保存。save 函数把 DataFrame 中的数据保存到文件或者用具体的格式来指明要读取的文件的类型,以及用具体的格式来指出要输出的文件是什么类型。

1. 保存操作结果

上一节的案例中读进来的数据中的 name 被筛选出来之后,可以通过 save 方法对此结果进行保存,具体操作如下:

```
scala > peopleDF.select("name").write.format("json").save("/examples/peopleresult.json")
……省略日志信息
16/04/30 16:52:22 INFO datasources.DefaultWriterContainer: Job job_201604301652_0000 committed.
16/04/30 16:52:22 INFOjson.JSONRelation: Listing hdfs://Master:9000/examples/peopleresult.json on driver
```

2. 查看保存结果

保存以后的结果,同样可以通过 HDFS 的 Web 控制台界面查看输出的结果,如图 3-2 所示。

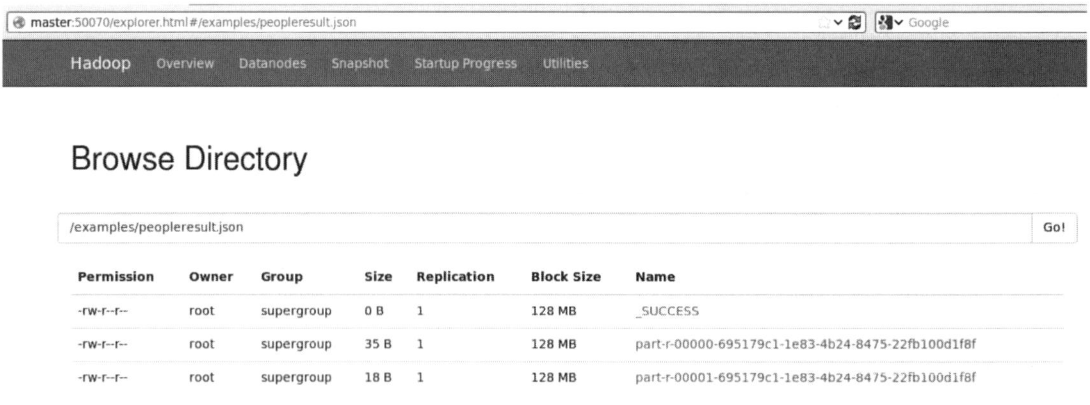

图 3-2 Web 端显示成功

3.1.3 综合案例——电商热销商品排名

本节介绍的案例是要实现对电商热销商品的排名操作。在该案例中,将演示如何通过

第3章　Spark SQL操作多种数据源

Spark SQL 来读取和存储数据，以不同方式创建"商品销售数据"和"商品价格"两个 DataFrame 对象，把 DataFrame 对象与 RDD 进行转换，然后进行 Join 操作，动态创建了新的 DataFrame 对象，最后输出新 DataFrame 对象中的数据。

1. 准备数据

在 D 盘根目录下，创建商品数据文件 goods.json，包括 3 条记录，包括商品名称和数量，内容如下：

```
{"name":"Bag","num":98}
{"name":"iPhone","num":100}
{"name":"Book","num":58}
```

2. 代码实现

基于前两节介绍的，通过 Spark SQL 读取和存储数据的方法实现本案例的代码如下：

```java
import java.util.ArrayList;
import java.util.List;

import org.apache.spark.SparkConf;
import org.apache.spark.api.java.JavaPairRDD;
import org.apache.spark.api.java.JavaRDD;
import org.apache.spark.api.java.JavaSparkContext;
import org.apache.spark.api.java.function.Function;
import org.apache.spark.api.java.function.PairFunction;
import org.apache.spark.sql.DataFrame;
import org.apache.spark.sql.Row;
import org.apache.spark.sql.RowFactory;
import org.apache.spark.sql.SQLContext;
import org.apache.spark.sql.types.DataTypes;
import org.apache.spark.sql.types.StructField;
import org.apache.spark.sql.types.StructType;
import scala.Tuple2;

public class SparkSQLWithJoin {
    public static void main(String[] args) {
        SparkConf conf = new SparkConf().setMaster("local").setAppName("SparkSQLWithJoin");
        JavaSparkContext sc = new JavaSparkContext(conf);
        SQLContext sqlContext = new SQLContext(sc);

        /**
         * 读取商品销售数据 Json 文件,创建商品销售数据 DataFrame 对象
         */
```

```java
DataFrame goodsDF = sqlContext.read().json("E:\\goods.json");

/**
 * 基于商品销售数据 DataFrame 来注册临时表
 */
goodsDF.registerTempTable("goodsNum");

/**
 * 查询出销售数量大于 90 的商品
 */
DataFrame excellentNumDF = sqlContext.sql("select name,num from goodsNum where num >90");

/**
 * 在 DataFrame 的基础上转化成为 RDD,再通过 map 操作获得销售数量大于 90 的所有商品的名称
 */
List<String> excellentNumNameList = excellentNumDF.javaRDD().map(new Function<Row,String>(){

    public String call(Row row) throws Exception {
        // TODO Auto-generated method stub
        return row.getAs("name");
    }

}).collect();

/**
 * 动态构建商品价格 Json 数据
 */
List<String> goodsPrice = new ArrayList<String>();
goodsPrice.add("{\"name\":\"Bag\",\"price\":200}");
goodsPrice.add("{\"name\":\"iPhone\",\"price\":5000}");
goodsPrice.add("{\"name\":\"Book\",\"price\":30}");

/**
 * 将上述 List 集合变成 RDD
 */
JavaRDD<String> goodsPriceRDD = sc.parallelize(goodsPrice);

/**
 * 再通过 RDD 来构建商品价格 DataFrame 对象
 */
```

第3章　Spark SQL操作多种数据源

```java
         */
        DataFrame goodsPriceDF = sqlContext.read().json(goodsPriceRDD);

        /**
         * 把商品价格 DataFrame 对象注册为临时表
         */
        goodsPriceDF.registerTempTable("goodsPrice");

        //遍历 excellentNumNameList 集合，组装 SQL 语句
        String sqlText = "select name,price from goodsPrice where name in(";
        for(int i = 0;i < excellentNumNameList.size();i++){
            sqlText += "'" + excellentNumNameList.get(i) + "'";
            if(i < excellentNumNameList.size() - 1){
                sqlText += ",";
            }
        }
        sqlText += ")";

        //执行该 SQL 语句,生成销售数量大于 90 的商品价格 DataFrame 对象
        DataFrame excellentNamePriceDF = sqlContext.sql(sqlText);

        /**
         * 将两个 DataFrame 通过 mapToPair 转化成新的 JavaRDD,再执行 join 操作
         */
        JavaPairRDD<String,Tuple2<Integer,Integer>> resultRDD = excellentNumDF.javaRDD()
        .mapToPair(new PairFunction<Row,String,Integer>(){

            private static final long serialVersionUID = 1L;

            @Override
            public Tuple2<String,Integer> call(Row row)throws Exception{

                return new Tuple2<String,Integer>((String)row.getAs("name"),(int)row.getLong(1));
            }
        }).join(excellentNamePriceDF.javaRDD().mapToPair(new PairFunction<Row,String,Integer>(){

            private static final long serialVersionUID = 1L;

            @Override
            public Tuple2<String,Integer> call(Row row)throws Exception{
```

```java
                    return new Tuple2<String,Integer>((String)row.getAs("name"),(int)row.getLong(1));
                }
            }));

/**
 * 转换 resultRDD 的类型为 JavaRDD<Row>
 */
JavaRDD<Row> resultRowRDD = resultRDD.map(new Function<Tuple2<String,Tuple2<Integer,Integer>>,Row>(){
    @Override
    public Row call(Tuple2<String,Tuple2<Integer,Integer>> tuple) throws Exception{
        //返回一行一行的内容
        return RowFactory.create(tuple._1,tuple._2._1,tuple._2._2);
    }
});

/**
 * 动态组拼元数据
 */
List<StructField> structFields = new ArrayList<StructField>();
structFields.add(DataTypes.createStructField("name",DataTypes.StringType,true));
structFields.add(DataTypes.createStructField("num",DataTypes.IntegerType,true));
structFields.add(DataTypes.createStructField("price",DataTypes.IntegerType,true));

//构建 StructType,用于最后 DataFrame 元数据的描述
StructType structType = DataTypes.createStructType(structFields);

DataFrame personsDF = sqlContext.createDataFrame(resultRowRDD,structType);

personsDF.show();

/**
 * 将处理的数据保存在 D 盘根目录下的 goodsResult 文件下
 */
personsDF.write().format("json").save("D:\\goodsResult");
    }
}
```

3. 运行结果

以上代码的运行结果如下,如期实现并输出了符合条件的新 DataFrame 对象中的数据:

第3章 Spark SQL操作多种数据源

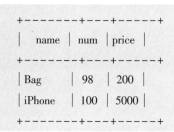

Section 3.2 Spark SQL 操作 Hive 示例

在目前企业级 Spark 大数据开发中，大多数情况下都是采用 Hive 来作为数据仓库的。Spark 提供了对 Hive 的支持，Spark 通过 HiveContext 可以直接操作 Hive 中的数据。基于 HiveContext，我们可以使用 sql/hql 两种方式来编写 SQL 语句对 Hive 进行操作，包括：创建表、删除表、往表中导入数据，以及对表中的数据进行 CRUD（增、删、改、查）操作。下面就开始动手实战。

本案例使用 Scala 语言开发，在 Spark 中使用 Hive 数据库，通过 HiveContext 使用 Join 基于 Hive 中的两张表（人员信息表、人员分数表）进行关联，查询大于90分的人的姓名、分数、年龄。演示了对 Hive 的常用操作（例如删除表、新建表、加载表数据、保存表数据），然后打包递交到 Spark 集群中运行。具体实现如下：

1. 准备数据

在 /home/Document/resource 目录下，创建两个文件：people.txt 和 peoplescores.txt，people.txt 文件是人员信息表，包括人员姓名和年龄信息；Peoplescores.txt 是人员分数表，包括人员姓名和分数信息。

Michael	29
Andy	30
Justin	19

Peoplescorees.txt 的文件内容如下：

Michael	99
Andy	97
Justin	68

2. 代码实现

以下代码实现在 Hive 中新建表以及加载表数据，使用 Join 将人员信息表、人员分数表进行关联，查询大于90分的人的姓名、分数、年龄。

```
package com.dt.spark.sql

import org.apache.spark.SparkConf
```

```
import org.apache.spark.SparkContext
import org.apache.spark.sql.hive.HiveContext

/**
 * @author Jonson
 */
object SparkSQLOnHive {
  def main(args:Array[String]):Unit = {
    val conf = newSparkConf().setMaster("spark://Master:7077").setAppName("SparkSQLOnHive")
    val sc = newSparkContext(conf)

    /**
     * 第一:直接通过 saveAsTable 的方式把 DataFrame 的数据保存到 Hive 数据仓库中
     * 第二:可以直接通过 HiveContext.table 方法来直接加载 Hive 中的表而生成 DataFrame
     */
    val hiveContext = new HiveContext(sc)

    //使用 Hive 数据仓库中的 Hive 数据库
    hiveContext.sql("use hive")                                      //需要提前在 Hive 中创建 Hive 表
    hiveContext.sql("DROP TABLE IF EXISTS people")                   //删除同名表
    hiveContext.sql("CREATE TABLE IF NOT EXISTS people(name INT,age INT)")  //创建自定义的表

    /**
     * 把本地数据加载到 Hive 数据仓库中,背后实际上发生了数据的复制
     * 当然,也可以通过 LOAD DATA INPATH 获得 HDFS 等上面的数据到 Hive 中(此时发生了
     数据的移动)
     */
    hiveContext.sql("LOAD DATA LOCAL INPATH '/home/Document/resource/people.txt 'INTO
       TABLE people")

    hiveContext.sql("DROP TABLE IF EXISTS peoplescores")  //删除 peoplescores 表
    hiveContext.sql("CREATE TABLE IF NOT EXISTS peoplescores(name INT,score INT)")
    hiveContext.sql("LOAD DATA LOCAL INPATH '/home/Document/resource/peoplescores.txt '
       INTO TABLE peoplescores")

    /**
     * 通过 HiveContext 使用 join 直接基于 Hive 中的两张表进行操作,获得大于 90 分的人的
     name,score,age
     */
    val resultDF = hiveContext.sql("SELECT pi.name,pi.age,ps.score FROM people pi JOIN peo-
       plescores ps ON pi.name = ps.name" +
```

第3章　Spark SQL操作多种数据源

```
            "WHERE ps.score >90")

        /**
```

通过 saveAsTable 创建一张 Hive Managed Table，表（peopleinformationresult），peopleinformationresult 表数据的元数据和数据即将放的位置都是由 Hive 数据仓库进行管理的，当删除该表的时候，数据也会一起被删除（磁盘上的数据不再存在）。

```
     * resultDF 是大于90分的人员信息表 Data Frame，包括姓名、分数，调用 Data Frame 的 saveAsTable 方法,将 resultDF 中大于90分的记录保存到 Hive 数据库表 peopleinformationresult 中
     */
    hiveContext.sql("DROP TABLE IF EXISTS peopleinformationresult")
    resultDF.saveAsTable("peopleinformationresult")

        /**
         * 使用 HiveContext 的 table 方法可以直接读取 Hive 数据仓库中的 Table 并生成 DataFrame
         */
        val dataFrameHive = hiveContext.table("peopleinformationresult")
    dataFrameHive.show()
    }
}
```

3. 打成 JAR 包

之前我们在 Eclipse 中进行了编码，在本地运行通过后，需要将代码打成 JAR 包，然后将 JAR 包上传到 Spark 集群中运行。

单击鼠标右键，选择 export 命令，选择 JAR file，然后点击 Next 按钮，如图 3-3 所示。选择 Browse 按钮指定 JAR 导出的路径，如图 3-4 所示。

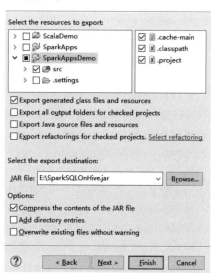

图 3-3　选择 JAR file　　　　图 3-4　指定 JAR 文件导出的路径

将 SparkSQL OnHive. JAR 包复制到/Home/Document/SparkApps/目录下，如图 3-5 所示。

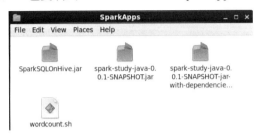

图 3-5 复制 JAR 文件

4. 编辑脚本并运行代码

在 Linux 系统中运行 Spark – Submit 时，如果不编写脚本，每次都要在 Linux 提示符中输入 Spark – Submit 运行的参数及相应的 JAR 包，调测不是很方便。因此，我们编写脚本文件，将 Spark – Submit 的相关内容编写在 wordcount.sh 脚本文件中，wordcount.sh 脚本可以放在 Linux 自定义的目录下（例如/usr/local/wordcount/），并且通过 chmod u + x wordcount.sh 赋予 wordcount.sh 可执行权限。这样运行 Spark 应用程序，每次只要运行 wordcount.sh 就可以了。

编辑 wordcount.sh 脚本如下：

```
/usr/local/spark/spark – 1.6.3 – bin – hadoop2.6/bin/spark – submit
– – class com.dt.spark.sql.SparkSQLOnHive – – file /usr/local/hive/hive – 1.2.1 – bin/conf/hive – site.xml
– – driver – class – path /usr/local/hive/apache – hive – 1.2.1 – bin/mysql – connector – java – 5.1.35 – bin.jar
– – master spark://Master:7077 /root/home/Document/SparkApps/SparkSQLOnHive.jar
```

启动 Hive 的 metastore 步骤如下：

```
[root@ Master SparkApps]# hive – – servicemetastore&
[1] 6028
[root@ Master SparkApps]# SLF4J:Class path contains multiple SLF4J bindings.
SLF4J:Found binding in [jar:file:/usr/local/hadoop/hadoop – 2.6.0/share/hadoop/common/lib/slf4j – log4j12-1.7.5.jar!/org/slf4j/impl/StaticLoggerBinder.class]
SLF4J:Found binding in [jar:file:/usr/local/spark/spark – 1.6.3 – bin – hadoop2.6/lib/spark – assembly – 1.6.0 – hadoop2.6.0.jar!/org/slf4j/impl/StaticLoggerBinder.class]
SLF4J:See http://www.slf4j.org/codes.html#multiple_bindings for an explanation.
SLF4J:Actual binding is of type [org.slf4j.impl.Log4jLoggerFactory]
Starting HiveMetastore Server
[root@ Master SparkApps]# SLF4J:Class path contains multiple SLF4J bindings.
SLF4J:Found binding in [jar:file:/usr/local/hadoop/hadoop – 2.6.0/share/hadoop/common/lib/slf4j – log4j12-1.7.5.jar!/org/slf4j/impl/StaticLoggerBinder.class]
SLF4J:Found binding in [jar:file:/usr/local/spark/spark – 1.6.3 – bin – hadoop2.6/lib/spark – assembly – 1.6.0 – hadoop2.6.0.jar!/org/slf4j/impl/StaticLoggerBinder.class]
SLF4J:See http://www.slf4j.org/codes.html#multiple_bindings for an explanation.
SLF4J:Actual binding is of type [org.slf4j.impl.Log4jLoggerFactory]
```

第3章 Spark SQL操作多种数据源

运行上述编辑过的脚本,命令行如下:

[root@ Master SparkApps]#./wordcount.sh

输出结果如下:

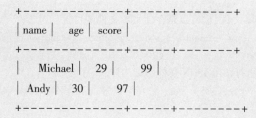

5. 查询执行结果

上述代码执行之后,可以在 Hive 中查询执行结果。各类查询命令行及结果如下:

```
hive > select * from people;              //查询人员信息:姓名、年龄
OK
Michael    29
Andy       30
Justin     19
Time taken:1.881 seconds,Fetched:3 row(s)
hive > select * frompeo plescores;        //查询人员分数信息:姓名,分数
OK
Michael    99
Andy       97
Justin     68
Time taken:0.335 seconds,Fetched:3 row(s)
Hive > select * from peopleinformationresult;   //查询大于90分的人的姓名、分数。这里的 Hive
数据库表 peopleinformationresult 是大于90分的人员信息记录,包括人名和分数。
OK
SLF4J:Failed to load class "org.slf4j.impl.StaticLoggerBinder".
SLF4J:Defaulting to no-operation(NOP) logger implementation
SLF4J:Seehttp://www.slf4j.org/codes.html#StaticLoggerBinder for further details.
Michael    99
Andy       97
Time taken:0.41 seconds,Fethed:2 row(s)
```

Section 3.3 Spark SQL 操作 JSON 数据集示例

Spark SQL 可以自动推断 JSON 数据集的模式,在 Spark 2.1.x 中,将其加载为 Dataset < Row >,可以使用 SparkSession.read.json() 读取 JSON 文件加载数据。

Spark SQL 操作 Json 数据集代码示例如下:

- 读入 JSON 文件加载 JSON 数据集。
- 读入 RDD[String] 加载 JSON 数据集，其中字符串 String 是 JSON 格式。

```java
import org.apache.spark.sql.Dataset;
import org.apache.spark.sql.Row;

//JSON 数据集指向路径可以是单个文本文件,也可以是存储文本文件的目录
Dataset<Row> people = spark.read().json("examples/src/main/resources/people.json");

//推断模式可以使用 printschema()方法打印出 JSON 文件的结构体
people.printSchema();
// root
//  |-- age: long (nullable = true)
//  |-- name: string (nullable = true)

//使用 DataFrame 创建临时视图
people.createOrReplaceTempView("people");

//SQL 语句可以通过使用 spark 提供的 sql 方法运行
Dataset<Row> namesDF = spark.sql("SELECT name FROM people WHERE age BETWEEN 13 AND 19");
namesDF.show();
// +------+
// | name|
// +------+
// |Justin|
// +------+

//或者,RDD[String] 存储每一个字符串的 JSON 对象,通过加载 JSON 数据集创建 DataFrame
List<String> jsonData = Arrays.asList(
        "{\"name\":\"Yin\",\"address\":{\"city\":\"Columbus\",\"state\":\"Ohio\"}}");
JavaRDD<String> anotherPeopleRDD =
        new JavaSparkContext(spark.sparkContext()).parallelize(jsonData);
Dataset anotherPeople = spark.read().json(anotherPeopleRDD);
anotherPeople.show();
// +---------------+----+
// |        address|name|
// +---------------+----+
// |[Columbus,Ohio]| Yin|
// +---------------+----+
```

Section 3.4 Spark SQL 操作 HBase 示例

Spark SQL 支持读取多种数据来源的数据，支持从 HBase 中读取数据，本节将实现从以

第3章 Spark SQL操作多种数据源

Java 的方式，通过 Spark 连接 hbaseTest 进行业务处理的案例：查询出 HBase 中满足设定的起始行至终止行条件的用户，统计各个级别用户的计数。

本节 Spark SQL 操作 HBase 案例中，HBase 数据库表名是"usertable"；info 是列族，是列的集合，一个列族中包含多个列，info 列族包括级别代码列（levelCode），以及其他的列；Row Key 是行键，记录每行的行 ID 值。

Row Key（行键）	Column Family（列族 info）	
	levelCode（级别代码）	其他列
195861 – 1035177490	levelCode1	……
……	……	……
195861 – 1072173147	levelCoden	……

Spark SQL 操作 HBase 案例具体实现步骤：

1）初始化 sparkContext，通过 jars 参数加载 Hbase 的 JAR 文件。
2）创建 HBase 的 Configuration 配置文件。
3）设置 Hbase 的查询条件。
Hbase Scan 的相关操作说明：
- setStartRow（byte[ ]startRow）：设置 Scan 的开始行。
- setStopRow（byte[ ] stopRow）：设置 Scan 的结束行 。
- addColumn（byte[ ] family，byte[ ] qualifier）：指定扫描的列。
- addFamily（byte[ ] family）：指定扫描的列族。
4）设置读取的 Hbase 表名。
5）获取获得 hbase 查询结果 Result。
6）从查询结果 result 中取出用户的等级，每次计数 1 次。
7）使用 reduceByKey 进行计数累加。
8）打印最终结果。

Spark SQL 操作 HBase 案例运行环境准备：

- 部署 Hadoop 集群。
- 部署 Hbase 集群。
- 部署 Spark 集群。

1. 代码实现

```
package com. dt. spark. streaming;

import org. apache. commons. logging. Log;
import org. apache. commons. logging. LogFactory;
import org. apache. hadoop. conf. Configuration;
import org. apache. hadoop. hbase. HBaseConfiguration;
import org. apache. hadoop. hbase. client. Result;
import org. apache. hadoop. hbase. client. Scan;
import org. apache. hadoop. hbase. io. ImmutableBytesWritable;
```

```java
import org.apache.hadoop.hbase.mapreduce.TableInputFormat;
import org.apache.hadoop.hbase.util.Base64;
import org.apache.hadoop.hbase.util.Bytes;
import org.apache.spark.api.java.JavaPairRDD;
import org.apache.spark.api.java.JavaSparkContext;
import org.apache.spark.api.java.function.Function2;
import org.apache.spark.api.java.function.PairFunction;
import scala.Tuple2;

import java.io.ByteArrayOutputStream;
import java.io.DataOutputStream;
import java.io.IOException;
import java.io.Serializable;
import java.util.List;

public class HbaseTest implements Serializable {

    public Log log = LogFactory.getLog(HbaseTest.class);

    static String convertScanToString(Scan scan) throws IOException {
        ByteArrayOutputStream out = new ByteArrayOutputStream();
        DataOutputStream dos = new DataOutputStream(out);
        scan.write(dos);
        return Base64.encodeBytes(out.toByteArray());
    }

    public void start() {
        //初始化 sparkContext,Spark SQL 连接 Hbase 需加载 HBase 的 JAR 包,
        //否则会报 unread block data 异常
        JavaSparkContext sc = new JavaSparkContext("spark://Master:7077","hbaseTest",
                "/usr/local/spark-1.6.0",
                new String[]{"target/ndspark.jar","target\\dependency\\hbase-0.94.6.jar"});

        //使用 HBaseConfiguration.create()生成 Configuration
        //必须在项目 classpath 下放上 Hadoop 及 HBase 的配置文件
        Configuration conf = HBaseConfiguration.create();
        //设置查询条件,这里的值返回用户的等级
        Scan scan = new Scan();
        scan.setStartRow(Bytes.toBytes("195861-1035177490"));
        scan.setStopRow(Bytes.toBytes("195861-1072173147"));
        scan.addFamily(Bytes.toBytes("info"));
```

```java
            scan.addColumn(Bytes.toBytes("info"),Bytes.toBytes("levelCode"));

        try{
            //需要读取的HBase表名
            StringtableName = "usertable";
            conf.set(TableInputFormat.INPUT_TABLE,tableName);
            conf.set(TableInputFormat.SCAN,convertScanToString(scan));

            //获得HBase查询结果Result
JavaPairRDD<ImmutableBytesWritable,Result> hBaseRDD = sc.newAPIHadoopRDD(conf,
TableInputFormat.class,ImmutableBytesWritable.class,
            Result.class);

            //从Result中取出用户的等级,并且每一个算一次
JavaPairRDD<Integer,Integer> levels = hBaseRDD.map(
            newPairFunction<Tuple2<ImmutableBytesWritable,Result>,Integer,Integer>(){
                @Override
                publicTuple2<Integer,Integer> call(
Tuple2<ImmutableBytesWritable,Result> immutableBytesWritableResultTuple2)
                    throws Exception{
                    byte[] o = immutableBytesWritableResultTuple2._2().getValue(
                        Bytes.toBytes("info"),Bytes.toBytes("levelCode"));
                    if(o != null){
                        return newTuple2<Integer,Integer>(Bytes.toInt(o),1);
                    }
                    return null;
                }
            });

            //数据累加
JavaPairRDD<Integer,Integer> counts = levels.reduceByKey(new Function2<Integer,
    Integer,Integer>(){
                public Integer call(Integer i1,Integer i2){
                    return i1 + i2;
                }
            });

            //打印出最终结果
List<Tuple2<Integer,Integer>> output = counts.collect();
for(Tuple2 tuple :output){
                System.out.println(tuple._1 + ":" + tuple._2);
            }
```

```
                    } catch( Exception e) {
                        log. warn(e) ;
                    }

                }

            /**
             * Spark 分布式计算时,如果类计算没写在 main 里面,实现的类必须继承 Serializable 接口,即
             HbaseTest 类必须继承序列化接口,HbaseTest implements Serializable 否则会报 Task not serializ-
             able: java. io. NotSerializableException 异常
                */
            public static void main( String[ ] args) throws InterruptedException {

                    newHbaseTest( ). start( ) ;

                    System. exit(0) ;
                }
        }
```

2. 输出结果

将上述代码打成 JAR 包,通过 SparkSubmit 向 Spark 集群提交运行,输出结果如下:

```
0:28528
11:708
4:28656
2:36315
6:23848
8:19802
10:6913
9:15988
3:31950
1:38872
7:21600
5:27190
12:17
```

Spark SQL 作为 Spark 大数据计算生态中一个重要的子项目,主要用来操作数据库,它提供了很多方法来操作不同的数据库,使得其可操作的数据来源变得非常广泛。Spark SQL 提供的大量 API 不仅易于使用,而且功能强大,不仅可以操作 Hive、HBase 等基于 HDFS (Hadoop Distributed File System) 的数据库,而且可以操作 MySQL、Oracle 等传统的关系型数据库,还可以操作 MongoDB 等 NoSQL 非关系型数据库,甚至可以操作 Kafka 分布式消息

系统。Spark SQL 支持多语言编程,包括 Java、Scala、Python 及 R。

以下我们采用 Spark SQL 分别操作 MySQL、MongoDB 数据库的内容,采用 2.10.4 版本的 Scala 语言开发,Spark 版本是 1.6.1,采用的操作系统是 64 位 Ubuntu kylin-15.10-desktop,集成开发环境是 IntelliJ IDEA。

3.5 Spark SQL 操作 MySQL 示例

MySQL 作为一种影响巨大的关系型数据库,在企业中广泛使用。本示例为在 Ubuntu 系统上安装 MySQL 数据库,并且以其作为数据来源用 Spark SQL 操作其数据。

3.5.1 安装并启动 MySQL

在 Ubuntu 系统上安装并启动 MySQL 数据库的方法,这里将 MySQL 服务器端及 MySQL 客户端都安装在本地。其过程如下:

1. 安装 MySQL 服务器端

在 Ubuntu 操作系统的命令终端输入 apt-get 命令,进行 MySQL 的服务器端安装,如下所示:

> apt-get install mysql-server

安装过程如下:

```
root@UbuntuDevelopment:~# apt-get install mysql-server
Reading package lists... Done
Building dependency tree
Reading state information... Done
The following extra packages will be installed:
  libhtml-template-perl mysql-server-5.6 mysql-server-core-5.6
Suggested packages:
  libipc-sharedcache-perl mailx tinyca
The following NEW packages will be installed:
  libhtml-template-perl mysql-server mysql-server-5.6 mysql-server-core-5.6
0 upgraded, 4 newly installed, 0 to remove and 234 not upgraded.
Need to get 10.3 MB of archives.
After this operation, 79.2 MB of additional disk space will be used.
Do you want to continue? [Y/n]
```

然后输入"Y",MySQL 服务器端安装完成。

2. 确认 MySQL 服务器端是否启动

用 netstat 命令查看 MySQL 的服务器端是否启动,命令如下:

> netstat -tap | grep mysql

若查看结果如下所示,则表明 MySQL 的服务器端安装并启动成功:

```
root@UbuntuDevelopment:~# netstat -tap | grep mysql
tcp        0      0 localhost:mysql         *:*                     LISTEN      4525/mysqld
root@UbuntuDevelopment:~#
```

3. 安装 MySQL 客户端

用 apt-get 命令执行 MySQL 的客户端的安装,如下所示:

```
> apt – get install mysql – client
```

安装过程如下：

```
root@UbuntuDevelopment:/usr/software/mysql# apt-get install mysql-client
Reading package lists... Done
Building dependency tree
Reading state information... Done
The following extra packages will be installed:
  libaio1 libdbd-mysql-perl libdbi-perl libmysqlclient18 libterm-readkey-perl
Suggested packages:
  libmldbm-perl libnet-daemon-perl libsql-statement-perl
The following NEW packages will be installed:
  libaio1 libdbd-mysql-perl libdbi-perl libmysqlclient18 libterm-readkey-perl
  mysql-common
0 upgraded, 9 newly installed, 0 to remove and 235 not upgraded.
Need to get 10.8 MB of archives.
After this operation, 74.9 MB of additional disk space will be used.
Do you want to continue? [Y/n]
```

输入"Y"，MySQL 客户端安装会成。

4. 启动 MySQL 的客户端

在命令终端用 mysql 的命令启动客户端，如下：

```
> mysql – h127.0.0.1 – u root – p ******
```

其中 –h 表示 MySQL 服务端主机，–u 表示 MySQL 的用户，–p 表示用户的密码。

```
Welcome to the MySQL monitor.    Commands end with ; or \g.
    Your MySQL connection id is 1 to server version: 4.0.16 – standard
    Type 'help;'or '\h 'for help. Type '\c 'to clear the buffer.
    mysql >
```

出现"mysql >"提示符，说明安装已经成功。

3.5.2 准备数据表

1. 建立数据库

在 MySQL 中建立一个数据库 MyDB，命令如下所示：

```
CREATE DATABASEMyDB;
SHOW DATABASES;
```

结果如下所示：

```
mysql> create database MyDB;
Query OK, 1 row affected (0.00 sec)

mysql> show databases;
+--------------------+
| Database           |
+--------------------+
| information_schema |
| MyDB               |
| mysql              |
| performance_schema |
+--------------------+
4 rows in set (0.00 sec)

mysql>
```

2. 建立表

在数据库 MyDB 中建立表 StudentInfo，表的字段如表 3-1 所示。

表 3-1 StudentInfo 的字段及其含义

字 段 名	类 型	含 义
ID	int（20）	学号
Name	varchar（20）	姓名
Gender	char（1）	性别
birthday	date	出生日期

命令如下：

```
> USE MyDB；
> createtable StudentInfo(
ID int(20) ,
Name varchar(20) ,
Gender char(1) ,
birthday date
);
```

3. 插入数据

用 INSERT INTO 命令向表 StudentInfo 中插入 5 条数据，如表 3-2 所示。

表 3-2 StudentInfo 数据

学 号	姓 名	性 别	出 生 日 期
1	A	F	1996 - 09 - 12
2	B	M	1995 - 12 - 23
3	C	M	1996 - 10 - 29
4	D	M	1995 - 02 - 25
5	E	F	1997 - 06 - 06

MySQL 命令如下：

```
> Insert into StudentInfo values(1,"A","F","1996-09-12");
> Insert into StudentInfo values(2,"B","M","1995-12-23");
> Insert into StudentInfo values(3,"C","M","1996-10-29");
> Insert into StudentInfo values(4,"D","M","1995-02-25");
> Insert into StudentInfo values(5,"E","F","1997-06-06");
```

4. 查看数据

查看表 StudentInfo 中的数据，命令如下：

```
SELECT * FROMStudentInfo;
```

结果如下：

```
mysql> select * from StudentInfo;
+----+------+--------+------------+
| ID | name | gender | birthday   |
+----+------+--------+------------+
|  1 | A    | F      | 1996-09-12 |
|  2 | B    | M      | 1995-12-23 |
|  3 | C    | M      | 1996-10-29 |
|  4 | D    | M      | 1995-02-25 |
|  5 | E    | F      | 1997-06-06 |
+----+------+--------+------------+
5 rows in set (0.00 sec)
```

5. 创建第二个表

再在数据库 MyDB 中创建第二个表 Score，其表结构如表 3-3 所示。

表 3-3 Score 的字段及其含义

字 段 名	类 型	含 义
ID	int（20）	学号
Name	varchar（20）	姓名
Score	float（10）	分数

在 MyDB 中创建表 Score 的命令如下：

```
Createtable Score(
ID int(20),
name varchar(20),
score float(10)
)
```

6. 插入数据

向表 Score 插入的数据，如表 3-4 所示。

表 3-4 Score 数据

学 号	姓 名	分 数
1	A	91
2	B	87
5	E	88
9	H	89
10	P	97

向表 Score 中插入数据的 MySQl 命令如下：

```
> Insert into Score values(1,"A",91);
> Insert into Score values(2,"B",87);
> Insert into Score values(5,"E",88);
> Insert into Score values(9,"H",89);
> Insert into Score values(10,"P",97);
```

7. 查看数据

查看表 Score 中的数据，命令如下：

```
SELECT * FROM Score;
```

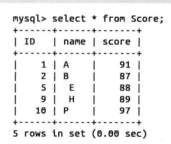

3.5.3 操作 MySQL 表

用 Spark SQL 操作 MySQL 数据库的表需要加载 mysql – connector – java 的 JAR 包 mysql – connector – java – 5.1.21.jar。这个 JAR 包是 Spark 连接 MySQL 的数据库驱动包,如已经下载,可以把该 JAR 包添加到项目的 Libraries 中;若还没有下载这个包,可以将 mySQL – connector – java 的依赖关系配置到 pom – xml 中,pom.xml 是 Naven 的 Jar 依赖关系管理配置文件,可通过 pom.xml 自动从网上下载这个 JAR 包。在 pom.xml 文件中添加 MySQL 的 JAR 包依赖如下操作:

```xml
<dependency>
<groupId>mysql</groupId>
<artifactId>mysql-connector-java</artifactId>
<version>5.1.35</version>
</dependency>
```

程序需要导入的类如下:

```scala
import java.sql.{Connection,DriverManager,Statement}
import org.apache.spark.sql.{DataFrameReader,Row,SQLContext}
import org.apache.spark.sql.types.{DataTypes,StructField,StructType}
import org.apache.spark.{SparkConf,SparkContext}
import scala.collection.mutable.ArrayBuffer
```

1. 主程序及相关代码

在项目中新建 Object,名字是 SparkSQLBookMysql,Spark 从 MySQL 数据库中读取学生信息表、分数表的信息,并进行相关的关联查询。主程序如下:

```scala
def main(args: Array[String]) {
    val conf = newSparkConf().setAppName("SparkSQLMysql")
    conf.setMaster("spark://localhost:7077") //程序在 Spark 集群运行,非本地运行
    //.setMaster("local[*]")
    val sc = newSparkContext(conf)
    sc.setLogLevel("WARN")
```

```
val sqlContext = new SQLContext(sc)

//jdbc 的 JAR 包可以放在 Spark 的 library 目录中,也可以在用 SparkSubmit 提交时指定 JAR 包的路径
//使用的 JAR 包是 mysql-connector-java-5.1.21.jar
val ip = "127.0.0.1"
val port = "3306"
var databaseName = "MyDB"  //数据库名
var userName = "root"  //用户名
var password = "123456"  //密码
var url = "jdbc:mysql://" + ip + ":" + port + "/" + databaseName

var studentInfoTable = "StudentInfo"  //学生信息表
var scoreTable = "Score"  //分数表

val mysqlDataFrameReader = sqlContext.read.format("jdbc")
  .option("url", url)  //数据库的 jdbc 链接地址
  .option("driver", "com.mysql.jdbc.Driver")  //驱动
  .option("user", userName)
  .option("password", password)

mysqlDataFrameReader.option("dbtable", studentInfoTable)
val studentInfoDataFrame = mysqlDataFrameReader.load()  //基于表 StudentInfo 生成的 DataFrame

//给表 studentInfoDataFrame 赋予 StructType,描述表的元数据,例如数据类型
val studentInfoStructed = setStudentInfoStructType(sqlContext, studentInfoDataFrame)
studentInfoStructed.show()
studentInfoStructed.select("ID", "Name").show()

mysqlDataFrameReader.option("dbtable", scoreTable)
val scoreDataFrame = mysqlDataFrameReader.load()  //基于表 Score 生成的 DataFrame

//给表 scoreDataFrame 赋予 StructType,描述表的元数据,例如数据类型
val scoreStructed = setScoreStructType(sqlContext, scoreDataFrame)
scoreStructed.show()
scoreStructed.select("ID", "Score").show()

studentGenderCount(studentInfoStructed)
studentRandomSplit(studentInfoStructed)
joinTables(sc, studentInfoStructed, scoreStructed)  //两个表做 inner join 操作
fileterScore(scoreStructed)
```

```
        sortScore(scoreStructed)
            scoreTopN(sqlContext,scoreStructed)
        getMaxScore( scoreStructed)
        getAverageScore(scoreStructed)
        increaseScore(sqlContext,scoreStructed)
            insertIntoStudentInfo(sqlContext)
            studentInfoDataFrame.show( )

        sc.stop( )
    }
```

代码说明：

1) setStudentInfoStructType 方法：第一个参数传入 sqlContext，第二个参数传入 studentInfoDataFrame，将 studentInfoDataFrame 进行 map 转换为 Rows 格式，通过 getStudentInfoStructType 构建学生信息的 schema；然后通过 SQLContext 的 createDataFrame（row，schema）方法创建 DataFrame，即结构化的学生信息 DataFrame（studentInfoStructed），基于学生信息 DataFrame（studentInfoStructed）可以统计学生 ID、学生姓名；性别分别为男、女的学生数量等信息。

2) setScoreStructType 方法：第一个参数传入 sqlContext，第二个参数传入 scoreDataFrame，将 scoreDataFrame 进行 map 转换为 Rows 格式，通过 getScoreSchema 构建分数信息的 schema；然后通过 SQLContext 的 createDataFrame（row，schema）方法创建 DataFrame，即结构化的分数信息 DataFrame（scoreStructed），基于分数信息 DataFrame（scoreStructed）可以统计分数大于 90 分的信息；分数从高到低排名查询；查询分数最高的学生的 ID 和姓名；分数小于 90 的分数加 5 分等操作。

```
    /**
      * 构建 studentInfoStructType,用于描述 loadSourceDataFrame 的元数据
      * @return
      */
    def getStudentInfoStructType( ) :StructType = {
        val studentInfoStructFields = newArrayBuffer[StructField]( )
            studentInfoStructFields. + = (DataTypes.createStructField( "ID", DataTypes.IntegerType,
true))
            studentInfoStructFields. + = (DataTypes.createStructField( "Name", DataTypes.StringType,
true))
            studentInfoStructFields. + = (DataTypes.createStructField( "Gender", DataTypes.StringType,
true))
            studentInfoStructFields. + = (DataTypes.createStructField( "Birthday", DataTypes.DateType,
true))
        StructType(studentInfoStructFields)
    }
```

```
/**
 * 构建 scoreStructFields,用于描述 loadSourceDataFrame 的元数据
 * @return
 */
def getScoreSchema() :StructType = {
    val scoreStructFields = new ArrayBuffer[StructField]()
    scoreStructFields. += (DataTypes. createStructField("ID", DataTypes. IntegerType, true))
    scoreStructFields. += (DataTypes. createStructField("Name", DataTypes. StringType, true))
    scoreStructFields. += (DataTypes. createStructField("Score", DataTypes. DoubleType, true))
    StructType(scoreStructFields)
}

/**
 * 给 StudentInfo 表添加元数据信息
 * @paramsqlContext
 * @paramdataFrame
 * @return
 */
def setStudentInfoStructType(sqlContext:SQLContext,dataFrame:DataFrame):DataFrame = {
    val rdd = dataFrame. map { x =>
        Row(x.getInt(0), x.getString(1), x.getString(2), x.getDate(3))
    }
    sqlContext. createDataFrame(rdd, getStudentInfoStructType)
}

/**
 * 给 score 表添加元数据信息
 * @paramsqlContext
 * @paramdataFrame
 * @return
 */
def setScoreStructType(sqlContext:SQLContext,dataFrame:DataFrame):DataFrame = {
    val rdd = dataFrame. map { x =>
        Row(x.getInt(0), x.getString(1), x.getDouble(2))
    }
    sqlContext. createDataFrame(rdd, getScoreSchema())
}
```

代码说明

1) getStudentInfoStructType 方法构建学生信息的 schema：创建一个元素类型为 StructField 的 ArrayBuffer 数组 studentInfoStructFields，其中 StructField 是一个 case class，case class 第一个成员变量是名称 name，第二个成员变量是数据类型 dataType，第三个成员变量是 nullable，

是否允许字段为空值；第四个成员变量是 metadata，是该字段的元数据，如果列的内容未被修改，在 selection 时元数据应保存，元数据类型如 metadata.jsonValue 等，metadata 默认设置为 Metadata.empty。在 studentInfoStructFields 中依次加入（ID、整型、允许空）、(姓名、字符串型、允许空)、(性别、字符串型、允许空)（生日、字符串型、允许空）的字段，返回 StructType（studentInfoStructFields）结构化 schema。

2) getScoreSchema 方法构建分数信息的 schema：创建一个元素类型为 StructField 的 ArrayBuffer 数组 scoreStructFields，在 scoreStructFields 中依次加入（ID、整型、允许空）、(姓名、字符串型、允许空)、(分数、浮点数型、允许空）的字段，返回 StructType（scoreStructFields）结构化 schema。

写完代码后，用 IntelliJ IDEA 将其打成 JAR 包，JAR 包的名字是 SparkSQLBOOK.jar，写脚本提交给 Spark 运行。具体的写脚本和打包过程在此不再评述，程序提交的脚本如下：

```
#! /bin/bash
./spark-submit --class SparkSQLBookMysql --master spark://127.0.0.1:7077 SparkSQLBOOK.jar
```

提交程序时把该脚本文件放到 Spark 的 bin 目录下。

2. 相关操作示例

下面是查询相关的系列示例。

1）示例一：查询表 StudentInfo 和 Score 中的内容。

本示例用于查询表 StudentInfo 和 Score 中的内容，代码如下：

```
/**
显示表的内容
@param sc
@parammysqlOptions
*/
def showTables(sc: SparkContext, mysqlOptions: DataFrameReader): Unit = {
  val sourceDataFrame = mysqlOptions.load()
  sourceDataFrame.show()
}
```

输出结果如下所示。

```
16/04/24 22:59:20 INFO TaskScheduler:
16/04/24 22:59:20 INFO DAGScheduler:
16/04/24 22:59:20 INFO DAGScheduler:
+---+----+------+----------+
| ID|Name|Gender|  Birthday|
+---+----+------+----------+
|  1|   A|     F|1996-09-12|
|  2|   B|     M|1995-12-23|
|  3|   C|     M|1996-10-29|
|  4|   D|     M|1995-02-25|
|  5|   E|     F|1997-06-06|
+---+----+------+----------+
16/04/24 22:59:20 INFO SparkContext:
16/04/24 22:59:20 INFO DAGScheduler:
```

StudentInfo 输出结果

```
16/04/24 22:59:20 INFO DAGScheduler:
16/04/24 22:59:20 INFO TaskScheduler:
16/04/24 22:59:20 INFO DAGScheduler:
+---+----+-----+
| ID|Name|Score|
+---+----+-----+
|  1|   A| 91.0|
|  2|   B| 87.0|
|  5|   E| 88.0|
|  9|   H| 89.0|
| 10|   P| 97.0|
+---+----+-----+
16/04/24 22:59:20 INFO SparkContext:
16/04/24 22:59:20 INFO DAGScheduler:
```

Score 输出结果

2）示例二：Select 部分字段。

```
studentInfoStructed.select("ID","Name").show()      //选择 ID,Name
scoreInfoStructed.select("ID","Score").show()       //选择 ID,Score
```

结果如下：

```
16/04/24 22:59:20 INFO DAGScheduler:        16/04/24 22:59:21 INFO DAGScheduler:
16/04/24 22:59:20 INFO DAGScheduler:        16/04/24 22:59:21 INFO DAGScheduler:
+---+----+                                  +---+-----+
| ID|Name|                                  | ID|Score|
+---+----+                                  +---+-----+
|  1|   A|                                  |  1| 91.0|
|  2|   B|                                  |  2| 87.0|
|  3|   C|                                  |  5| 88.0|
|  4|   D|                                  |  9| 89.0|
|  5|   E|                                  | 10| 97.0|
+---+----+                                  +---+-----+

16/04/24 22:59:20 INFO SparkContext:        16/04/24 22:59:21 INFO SparkContext:
16/04/24 22:59:20 INFO DAGScheduler:        16/04/24 22:59:21 INFO DAGScheduler:
16/04/24 22:59:20 INFO DAGScheduler:
```

　　　　选择 ID，Name 结果　　　　　　　　　　选择 ID，Score 结果

3）示例三：统计男女数量。

```
/**
    * 统计男女数量
    *
    * @param studentInfoStructed
    */
def studentGenderCount(studentInfoStructed: DataFrame): Unit = {
    studentInfoStructed.groupBy(studentInfoStructed("Gender")).count().show()
}
```

程序输出结果如下：

```
+------+-----+
|Gender|count|
+------+-----+
|     F|    2|
|     M|    3|
+------+-----+
```

4）示例四：随机分成两组。

```
/**
    * 随机分成两组
    * 在数据挖掘中把数据分成两组，一组用作训练数据，另一组用作测试数据
    * @param scoreStructed
    */
def studentRandomSplit(scoreStructed: DataFrame): Unit = {
    val weights =   Array[Double](0.7,0.3)
```

```
        val splited = scoreStructed.randomSplit(weights)
        splited(0).show()
        splited(1).show()
    }
```

程序输出结果如下:

```
+---+----+------+----------+
| ID|Name|Gender|  Birthday|
+---+----+------+----------+
|  3|   C|     M|1996-10-29|
|  4|   D|     M|1995-02-25|
|  5|   E|     F|1997-06-06|
+---+----+------+----------+

+---+----+------+----------+
| ID|Name|Gender|  Birthday|
+---+----+------+----------+
|  1|   A|     F|1996-09-12|
|  2|   B|     M|1995-12-23|
+---+----+------+----------+
```

5) 示例五:对表进行 inner join。

```
    /**
      * 对表进行 inner jion
      *
      * @param sc
      * @param studentInfoDataFrame
      * @paramscoreDataFrame
      */
    def joinTables(sc: SparkContext, studentInfoDataFrame: DataFrame, scoreDataFrame: DataFrame): Unit = {
        val joinColumns = "ID"
        val joinedDataFreme =  studentInfoDataFrame.join(scoreDataFrame, joinColumns)
        joinedDataFreme.show()
    }
```

程序输出结果如下:

```
+---+----+------+----------+----+-----+
| ID|name|gender|  birthday|Name|Score|
+---+----+------+----------+----+-----+
|  1|   A|     F|1996-09-12|   A| 91.0|
|  2|   B|     M|1995-12-23|   B| 87.0|
|  5|   E|     F|1997-06-06|   E| 88.0|
+---+----+------+----------+----+-----+
```

6) 示例六:取出 Score 大于 90 分的信息。

```
    /**
      * 取出 Score 大于 90 分的信息
```

```
       *
       * @paramscoreStructed
       */
     def fileterScore(scoreStructed:DataFrame):Unit = {
   scoreStructed.filter(scoreStructed("Score") >= 90).show()
     }
```

程序输出结果如下：

```
+---+----+-----+
| ID|Name|Score|
+---+----+-----+
|  1|   A| 91.0|
| 10|   P| 97.0|
+---+----+-----+
```

7）示例七：分数从高到低排名。

```
   /**分数从高到低排名
     * @paramscoreStructed
     */
     def sortScore(scoreStructed:DataFrame):Unit = {
   scoreStructed.sort(scoreStructed("Score").desc).show()
     }
```

程序输出结果如下：

```
+---+----+-----+
| ID|Name|Score|
+---+----+-----+
| 10|   P| 97.0|
|  1|   A| 91.0|
|  9|   H| 89.0|
|  5|   E| 88.0|
|  2|   B| 87.0|
+---+----+-----+
```

8）示例八：统计分数的 top 3。

```
   /**
     *统计分数的 top 3
     * @paramsqlContext
     * @paramscoreStructed
     */
     def scoreTopN(sqlContext:SQLContext,scoreStructed: DataFrame):Unit = {
       val topn = scoreStructed.sort(scoreStructed("Score").desc).take(3)
       val rdd = sqlContext.sparkContext.makeRDD(topn)
   sqlContext.createDataFrame(rdd, getScoreSchema).show()
     }
```

程序输出结果如下：

```
+---+----+-----+
| ID|Name|Score|
+---+----+-----+
| 10|   P| 97.0|
|  1|   A| 91.0|
|  9|   H| 89.0|
+---+----+-----+
```

9）示例九：查询最高分。

```
/**
 * 查询分数最高的 Student 的 ID 和姓名
 * @param scoreStructed
 */
def getMaxScore(scoreStructed: DataFrame): Unit = {
  scoreStructed.agg(Map("Score" -> "max")).show()
}
```

程序输出结果如下：

```
+---------+
|max(Score)|
+---------+
|     97.0|
+---------+
```

10）示例十：求平均分数。

```
/**
 * 求平均分数
 *
 * @param scoreStructed
 */
def getAverageScore(scoreStructed: DataFrame): Unit = {
  scoreStructed.agg(Map("Score" -> "avg")).show()
}
```

程序输出结果如下：

```
+---------+
|avg(Score)|
+---------+
|     90.4|
+---------+
```

11）示例十一：把 Score 小于 90 的分数加 5 分。

```
/**
 * 把 Score 小于 90 的分数加 5 分
 *
 * @param sqlContext
 */
```

```
def increaseScore(sqlContext:SQLContext,scoreStructed:DataFrame):Unit = {
    val increased = scoreStructed.map(row => {
        val score = row.getDouble(2)
        val incscore = if (score < 90) score + 5 else score
        Row(row.get(0), row.getString(1), incscore)
    })
    sqlContext.createDataFrame(increased, scoreStructed.schema).show()
}
```

程序输出结果如下：

```
+---+----+-----+
| ID|Name|Score|
+---+----+-----+
|  1|   A| 91.0|
|  2|   B| 92.0|
|  5|   E| 93.0|
|  9|   H| 94.0|
| 10|   P| 97.0|
+---+----+-----+
```

12）示例十二：向表 studentInfo 中插入数据。

```
/**
 * 向表 studentInfo 中插入数据
 *
 * @paramsqlContext
 */
def insertIntoStudentInfo(sqlContext:SQLContext):Unit = {
    val rowlist = new java.util.ArrayList[Row]()
    val dateFormat = new SimpleDateFormat("yyyy-MM-dd")
    val day1 = dateFormat.parse("1997-06-12")
    val day2 = dateFormat.parse("1998-07-23")
    val day3 = dateFormat.parse("1994-12-12")
rowlist.add( Row (30,"X","F",new java.sql.Date(day1.getYear,day1.getMonth,day1.getDay)))
rowlist.add( Row (31,"Y","F",new java.sql.Date(day2.getYear,day1.getMonth,day1.getDay)))
rowlist.add( Row (32,"Z","M",new java.sql.Date(day3.getYear,day1.getMonth,day1.getDay)))

    val studentInfoDataframe = sqlContext.createDataFrame(rowlist, getStudentInfoStructType)
    studentInfoDataframe.show()
    val connectionProperties = new Properties()
    connectionProperties.setProperty("user","root")
    connectionProperties.setProperty("password","123456")
    studentInfoDataframe.write.mode(SaveMode.Append).jdbc
("jdbc:mysql://localhost:3306/MyDB","StudentInfo",connectionProperties)
}
```

程序输出结果如下:

```
+---+----+------+----------+
| ID|Name|Gender|  Birthday|
+---+----+------+----------+
| 30|   X|     F|1997-06-04|
| 31|   Y|     F|1998-06-04|
| 32|   Z|     M|1994-06-04|
+---+----+------+----------+

+---+----+------+----------+
| ID|name|gender|  birthday|
+---+----+------+----------+
|  1|   A|     F|1996-09-12|
|  2|   B|     M|1995-12-23|
|  3|   C|     M|1996-10-29|
|  4|   D|     M|1995-02-25|
|  5|   E|     F|1997-06-06|
| 31|   Y|     F|1998-06-04|
| 30|   X|     F|1997-06-04|
| 32|   Z|     M|1994-06-04|
+---+----+------+----------+
```

3.6 Spark SQL 操作 MongoDB 示例

MongoDB 作为一种非常重要的非关系型数据库,弥补了传统关系型数据库的不足。本节介绍在 Ubuntu 系统上安装 MongoDB 数据库,并且以 MongoDB 作为数据来源用 SparkSQL 操作其数据的方法。

3.6.1 安装配置 MongoDB

安装配置 MongoDB,其过程如下:

1. 访问官网并下载 MongoDB

在 Ubuntu 的浏览器中打开 MongoDB 的官网,网址是:https://www.MongoDB.org。如图 3-6 所示。

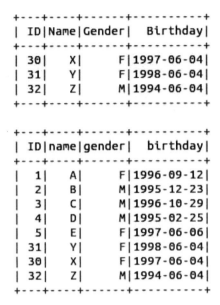

图 3-6 MongoDB 官网

单击 Download MongoDB 按钮（默认是 3.2 版本），进入下载页面。

选择操作系统类型和 MongoDB。这里选择的是 Ubuntu 14.04 Linux 64 – bit，如图 3 – 7 所示。

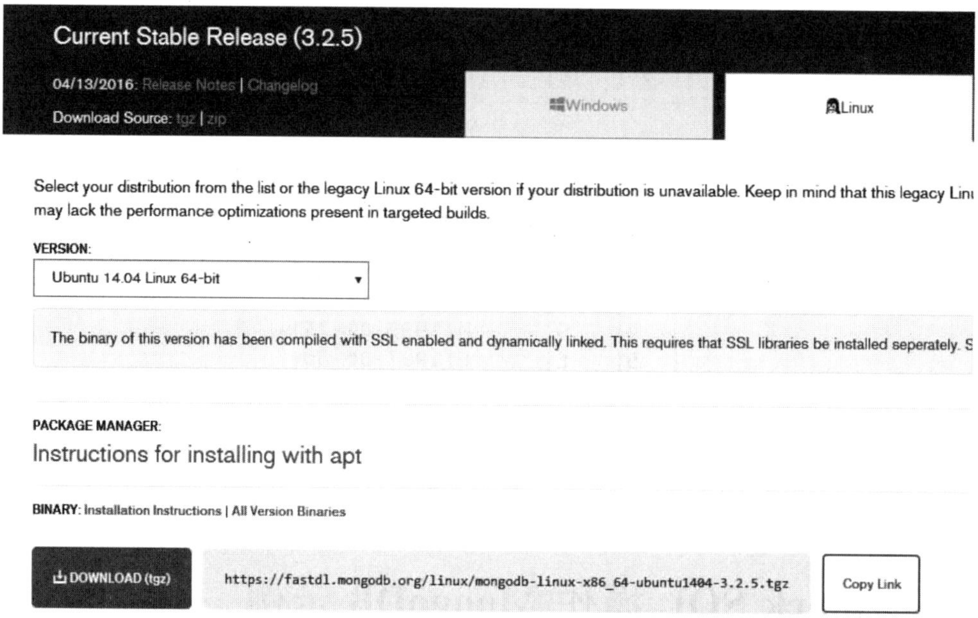

图 3 – 7　选择操作系统类型

单击 DOWNLOAD（tgz）按钮，开始下载，如图 3 – 8 所示。

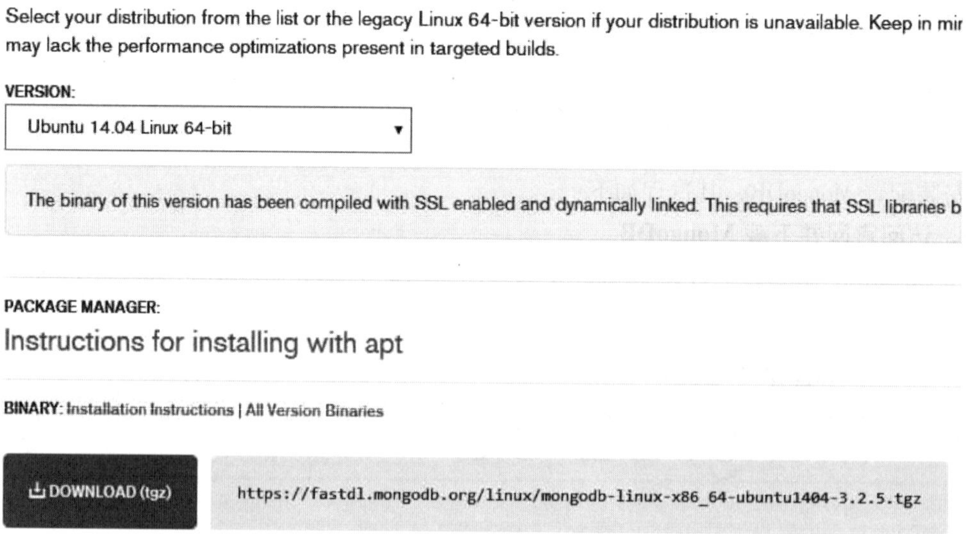

图 3 – 8　开始下载

2. 解压并安装 MongoDB

下载完毕后，用 tar 命令解压，文件解压到/usr/software/mongodb 目录下。

配置环境变量。用 vim 命令打开环境变量文件 ~/.bashrc，把解压后的目录配置到文件

~/.bashrc 中，如下所示：

```
export JAVA_HOME=/usr/software/java/jdk1.8.0_73
export JRE_HOME=${JAVA_HOME}/jre
export SCALA_HOME=/usr/software/scala/scala-2.10.4
export MONGODB_HOME=/usr/software/mongodb/mongodb-linux-x86_64-ubuntu1404-3.2.5

export CLASS_PATH=.:${JAVA_HOME}/lib:${JRE_HOME}/lib

export PATH=${MONGODB}/bin:${SCALA_HOME}/bin:${JAVA_HOME}/bin:$PATH
```

3.6.2 启动 MongoDB

Mongodb 是 NoSql 数据库，采用文档存储的存储方式。Mongodb 支持的查询语言非常强大，类似面向对象的查询语言，可以实现关系数据库单表查询的大部分功能，支持对数据建立索引。

Mongodb 的服务器部署方式包括：

- MongoDB 单机服务器部署：运行 MongoDB 包 bin 目录下的 mongod.exe，配置数据目录及日志文件，即可打开 MongoDB 服务。
- Mongodb 集群服务器部署：Mongodb 集群部署包括 Replica Set，Sharding，Master-Slaver 三种方式。

Mongodb 的客户端工具包括：

- Mongo Shell 客户端用来连接 MongoDB 的 JavaScript 接口，用户使用 Mongo Shell 查询和操作 MongoDB 中的数据、对 MongoDB 进行管理。
- Mongo VUE 工具提供一个简洁可用的 MongoDB 管理界面。
- Robomongo 是基于 Shell 的跨平台 MongoDB 可视化工具。
- MongoChef 是可视化的 Mongodb 数据库管理和查询工具。

本节 MongoDB 的讲解中，我们使用 MongoDB 单机服务器的部署方式，客户端使用 Mongo Shell 方式连接 MongoDB 服务器。

MongoDB 数据库启动步骤如下：

1. 创建数据目录和日志文件

打开一个命令终端，进入到 mondodb 目录，第一次启动时创建一个目录 data，用于存放 MongoDB 的数据。第一次启动时创建文件 log，用于保存日志。

2. 开启 MongoDB 服务端

进入到 bin 目录，开启 MongoDB 服务端。用 mongod 脚本开启服务端，如下所示：

```
>./mongod --dbpath ../data/ --logpath ../log --logappend
```

mongod --dbpath 创建数据库文件的存放位置，启动 mongodb 服务时需确定数据库文件存放的位置；--logpath 表示日志文件存放的路径；--logappend 表示以追加的方式写日志文件。

3. 开启 MongoDB 客户端

MongoDB Shell 是 MongoDB 自带的交互式 Javascript Shell 工具，对 MongoDB 数据库进行

操作和管理。进入 MongoDB 的 bin 目录，输入 ./mongo 命令进入 MongoDB 后台：

```
>./mongo
root@UbuntuDevelopment:/usr/software/mongodb/mongodb-linux-
mongodb-linux-x86_64-ubuntu1404-3.2.5/bin# ./mongo
MongoDB shell version: 3.2.5
connecting to: test
Server has startup warnings:
2016-04-19T22:18:58.579+0800 I CONTROL  [main] ** WARNING:
2016-04-19T22:18:58.579+0800 I CONTROL  [main] **
2016-04-19T22:18:58.668+0800 I CONTROL  [initandlisten]
2016-04-19T22:18:58.668+0800 I CONTROL  [initandlisten] **
nd_ip has been specified.
2016-04-19T22:18:58.668+0800 I CONTROL  [initandlisten] **
2016-04-19T22:18:58.668+0800 I CONTROL  [initandlisten] **
2016-04-19T22:18:58.668+0800 I CONTROL  [initandlisten] **
```

输入 ./mongo 命令进入 MongoDB 后台后，显示 MongoDB Shell 的版本号，默认链接到 test 文档（数据库），MongoDB Shell 是一个 JavaScript Shell，可以运行一些简单的算术运算：

```
>2+2
4
>3+6
9
```

3.6.3　准备数据

MongoDB 中基本的概念是文档、集合、数据库，如下表所示：MongoDB 中的 collection 对应于 SQL 数据库中的表；MongoDB 中的 document 对应于 SQL 数据库中的行；MongoDB 中的 field 对应于 SQL 数据库中的数据字段：

SQL 数据库	MongoDB	说　　明
database	database	数据库
table	collection	数据库表/集合
row	document	数据记录行/文档
column	field	数据字段/域
index	index	索引
table joins		表连接，MongoDB 不支持
primary key	primary key	主键，MongoDB 自动将_id 字段设置为主键

本节 MongoDB 案例的数据是学生信息表信息：包括 ID、姓名、性别、生日；

ID	Name（姓名）	Gender（性别）	Birthday（生日）
1	A	F	1996/9/12
2	B	M	1995/12/23
3	C	M	1996/10/29
4	D	M	1995/2/25
5	E	F	1997/6/6

第3章　Spark SQL操作多种数据源

MongoDB 文档是一组键值对（Key – Value）。MongoDB 的文档不需要设置相同的字段，相同的字段不需要相同的数据类型，这与关系型数据库有很大的区别。这里学生信息表信息中还包括学生的成绩等信息：ID、姓名、信息（性别、生日）、分数（数学、历史、汉语）。

ID	Name（姓名）	Info（信息）		Score（分数）		
		Gender(性别)	Birthday(生日)	Math(数学)	History(历史)	Chinese(汉语)
24	X	F	1997/6/6	90	87	
25	Y	M	1998/12/21	78	96	
26	Z	M	1998/3/21	83		84

我们在 MongoDB 数据库中创建学生信息表信息的相关记录：

在 MongoDB 的命令行客户端中，在 MongDB 中创建数据库 MyDB，命令如下：

> use MyDB;

向 MyDB 中插入一个集合 MyCollection，命令如下：

> db. createCollection("MyCollection");

向集合 MyCollection 中插入以下数据记录，命令如下：

db. MyCollection. save({ID:"1",Name:"A",Gender:"F",Birthday:"1996 – 09 – 12"});
db. MyCollection. save({ID:"2",Name:"B",Gender:"M",Birthday:"1995 – 12 – 23"});
db. MyCollection. save({ID:"3",Name:"C",Gender:"M",Birthday:"1996 – 10 – 29"});
db. MyCollection. save({ID:"4",Name:"D",Gender:"M",Birthday:"1995 – 02 – 25"});
db. MyCollection. save({ID:"5",Name:"E",Gender:"F",Birthday:"1997 – 06 – 06"});
db. MyCollection. save({ID:"24", Name:"X", Info:{Gender:"F", Birthday:"1997 – 06 – 06"}, Score:{Math:90,History:87}});
db. MyCollection. save({ID:"25", Name:"Y", Info:{Gender:"M", Birthday:"1998 – 12 – 21"}, Score:{Math:78,History:96}});
db. MyCollection. save({ID:"26", Name:"Z", Info:{Gender:"M", Birthday:"1998 – 03 – 21"}, Score:{Math:83,Chinese:84}});

用 find 命令查看数据结果，命令如下：

> db. MyCollection. find();

数据结果如下所示：

```
> db.MyCollection.find()
{ "_id" : ObjectId("5720e8dd6d110e7ec14b671a"), "ID" : "1", "Name" : "A", "Gender" : "F", "Birthday" : "1996-09-12" }
{ "_id" : ObjectId("5720e8f46d110e7ec14b671b"), "ID" : "2", "Name" : "B", "Gender" : "M", "Birthday" : "1995-12-23" }
{ "_id" : ObjectId("5720e8fb6d110e7ec14b671c"), "ID" : "3", "Name" : "C", "Gender" : "M", "Birthday" : "1996-10-29" }
{ "_id" : ObjectId("5720e9006d110e7ec14b671d"), "ID" : "4", "Name" : "D", "Gender" : "M", "Birthday" : "1995-02-25" }
{ "_id" : ObjectId("5720e9066d110e7ec14b671e"), "ID" : "5", "Name" : "E", "Gender" : "F", "Birthday" : "1997-06-06" }
{ "_id" : ObjectId("5720e9176d110e7ec14b671f"), "ID" : "24", "Name" : "X", "Info" : { "Gender" : "F", "Birthday" : "1997-06-06" }, "Score" : { "Math" : 90, "History" : 87 } }
{ "_id" : ObjectId("5720e9276d110e7ec14b6720"), "ID" : "25", "Name" : "Y", "Info" : { "Gender" : "M", "Birthday" : "1998-12-21" }, "Score" : { "Math" : 78, "History" : 96 } }
{ "_id" : ObjectId("5720e92e6d110e7ec14b6721"), "ID" : "26", "Name" : "Z", "Info" : { "Gender" : "M", "Birthday" : "1998-03-21" }, "Score" : { "Math" : 83, "Chinese" : 84 } }
```

3.6.4　Spark SQL 操作 MongoDB

本节 Spark SQL 操作 MongoDB 案例将对学生信息表的相关记录进行查询统计操作。具体步骤如下：

1）MongoDB 服务器启动 MongoDB 服务。
2）Spark 中引入 MongoDB 的相关 JAR 包。
3）设置 MongoDB 数据库连接的 URI 信息：地址、端口、数据库及文档信息。
4）设置 MongoDB 的查询条件。
5）Spark 代码打成 JAR 包，提交集群运行。Spark 从 MongoDB 中查询统计学生信息的数据。
- 查询性别为男的所有学生。
- 查询性别为男、数学成绩高于 80 分的文档。
- 数学成绩低于 90 的分数加上 5 分成绩。
- 删除历史 History 分数小于 90 的键。
- 查询结果保存到 MongoDB 数据库中。

Spark 中读取 MongoDB 数据库，需要依赖的 JAR 包包括：MongoDB – driver – 3.0.2.jar，MongoDB – driver – core – 3.0.2.jar，bson – 3.0.2.jar，mongo – hadoop – core – 1.4.0.jar，mongo – java – driver – 3.0.2.jar，spark – MongoDB_2.10 – 0.10.1.jar，可以在 Maven 的配置文件 pom.xml 中增加以下依赖关系：

```
<dependency>
    <groupId>org.mongodb</groupId>
    <artifactId>mongodb-driver</artifactId>
    <version>3.0.2</version>
</dependency>
<dependency>
    <groupId>org.mongodb</groupId>
    <artifactId>mongodb-driver-core</artifactId>
    <version>3.0.2</version>
</dependency>
<dependency>
    <groupId>org.mongodb</groupId>
    <artifactId>bson</artifactId>
    <version>3.0.2</version>
</dependency>
<dependency>
    <groupId>org.mongodb.mongo-hadoop</groupId>
    <artifactId>mongo-hadoop-core</artifactId>
    <version>1.4.0</version>
</dependency>
<dependency>
```

第3章 Spark SQL操作多种数据源

```
<groupId>org.mongodb</groupId>
<artifactId>mongo-java-driver</artifactId>
<version>3.0.2</version>
</dependency>
```

其中 mongo-hadoop-core-1.4.0.jar 是连接器，可以用它来实现从 MongoDB 上读写数据，其配置参数使用配置对象传递，其中最重要的两个参数是 mongo.input.uri 和 mongo.output.uri，这两个参数提供了 MongoDB 主机、端口、权限、数据库和数据集合名字。

在用 Spark 操作 MongoDB 的过程中，首先使用命令./mongod --dbpath ../data/ --logpath ../log ---logappend 启动 MongoDB 的服务。

然后用 Spark 操作 MngoDB 需要引入如下的类（org.bson.BasicBSONObject、org.bson.BSONObject、com.mongodb.hadoop.MongoInputFormat、com.mongodb.hadoop.MongoOutputFormat、org.apache.hadoop.conf.Configuration）

```
import org.apache.spark.{SparkContext,SparkConf}
import org.bson.{BasicBSONObject,BSONObject}
import com.mongodb.hadoop.{MongoInputFormat,MongoOutputFormat}
import org.apache.hadoop.conf.Configuration
```

引入 MongoDB 的 JAR 包以后，在 main 主程序中封装各方法操作 MongoDB，步骤如下：

- 初始化 Spark Context。
- 在主程序业务代码中分别调用各方法：queryDocuments（sc）、querySubcollection（sc）、updateMath（sc）、removeHistory（sc）、saveToMongo（sc）
- 关闭 Spark Context。

main 主函数代码如下：

```
/**
 * 主函数
 *
 * @param args
 */
def main(args: Array[String]) {val sparkConf = new SparkConf().setAppName("SparkMongoDB")
.setMaster("local[*]")//.setSparkHome(sys.env("SPARK_HOME"))
  val sc = new SparkContext(sparkConf)
  sc.setLogLevel("WARN")
queryDocuments(sc)
querySubcollection(sc)
updateMath(sc)
removeHistory(sc)
saveToMongo(sc)

  sc.stop()
}
```

（1）示例一：查询性别为男的所有学生

1）创建 Hadoop 的 Configuration 配置类，设置 MongoDB 的输入 input URI 连接属性：地址、端口、MyDB 数据库及 MyCollection 集合信息；设置 MongoDB 的 input 输入类型为 MongoInputFormat；设置 MongoDB 的查询条件 mongo.input.query：性别是男。

2）调用 SparkContext 的 newAPIHadoopRDD 方法，newAPIHadoopRDD 的第一个参数 Configuration 用于设置数据集的配置，Configuration 将被放进 Spark 广播中。这里传入 MongoDB 的配置类；第二个参数 InputFormat 为输入类型是 MongoInputFormat 格式；第三个参数是返回结果的 Key 值，类型为 Object；第四个参数是返回结果的 Value 值，类型为 BSONObject；

3）对 newAPIHadoopRDD 查询 MongoDB 的结果遍历，打印输出。第一个元素是 Object ID，第二个元素是 MongoDB 中 MyDB 数据库 MyCollection 集合中性别为男的文档 document 记录。

```
/**
 * 查询 Gender:M 的所有学生
 */
def queryDocuments(sc:SparkContext): Unit = {
    val mongoConfig = new Configuration()
mongoConfig.set("mongo.input.uri","mongodb://127.0.0.1:27017/MyDB.MyCollection")
mongoConfig.set("mongo.job.input.format","com.mongodb.hadoop.MongoInputFormat")
mongoConfig.set("mongo.input.query","{'Gender':'M'}")
    val mongoRDD = sc.newAPIHadoopRDD(mongoConfig,classOf[MongoInputFormat],classOf[Object],classOf[BSONObject])
mongoRDD.foreach(x => println(x._1 + " " + x._2))
}
```

查询结果如下：

```
5720e8f46d110e7ec14b671b { "_id" : { "$oid" : "5720e8f46d110e7ec14b671b" } , "ID" : "2" , "Name" : "B" , "Gender" : "M" , "Birthday" : "1995-12-23"}
5720e8fb6d110e7ec14b671c { "_id" : { "$oid" : "5720e8fb6d110e7ec14b671c" } , "ID" : "3" , "Name" : "C" , "Gender" : "M" , "Birthday" : "1996-10-29"}
5720e9006d110e7ec14b671d { "_id" : { "$oid" : "5720e9006d110e7ec14b671d" } , "ID" : "4" , "Name" : "D" , "Gender" : "M" , "Birthday" : "1995-02-25"}
```

（2）示例二：查询性别是男并且数学分数大于 80 分的文档

1）创建 Hadoop 的 Configuration 配置类，设置 MongoDB 的输入 input URI 连接属性：地址、端口、MyDB 数据库及 MyCollection 集合信息；设置 MongoDB 的 input 输入类型为 MongoInputFormat；设置 MongoDB 的查询条件 mongo.input.query：信息 Info 性别为男，分数 Score 为数学成绩大于 80 分。

2）调用 SparkContext 的 newAPIHadoopRDD 方法。

3）对 newAPIHadoopRDD 查询 MongoDB 的结果遍历，打印输出。第一个元素是 Object ID，第二个元素是 MongoDB 中 MyDB 数据库 MyCollection 集合中性别是男并且数学大于 80 分的记录。

```
/**
 * 读取 mongodb 中的内嵌文档
```

第3章 Spark SQL操作多种数据源

```
 * 查询 Gender 是 M 并且 Math 大于 80 的文档
 */
def querySubcollection(sc:SparkContext): Unit = {
    val mongoConfig = new Configuration()
mongoConfig.set("mongo.input.uri","mongodb://127.0.0.1:27017/MyDB.MyCollection")
mongoConfig.set("mongo.job.input.format","com.mongodb.hadoop.MongoInputFormat")
mongoConfig.set("mongo.input.query","{'Info':{$exists:true},'Info.Gender':'M',Score:{$exists:
true},Score.Math:{$gte:80}},{}")

    val mongoRDD = sc.newAPIHadoopRDD(mongoConfig,classOf[MongoInputFormat],classOf[Object],classOf[BSONObject])
mongoRDD.foreach(x = > println(x._1 + " " + x._2))
}
```

查询结果如下：

```
5720e92e6d110e7ec14b6721 { "_id" : { "$oid" : "5720e92e6d110e7ec14b6721" }, "ID" : "26", "Name" : "Z", "Info" : { "Gender" : "M",
"Birthday" : "1998-03-21" }, "Score" : { "Math" : 83.0, "Chinese" : 84.0}}
root@Ubuntu5:/usr/software/spark/spark-1.6.1-bin-hadoop2.6/bin#
```

（3）示例三：数学小于 90 分的分数加 5 分

1）创建 Hadoop 的 Configuration 配置类，设置 MongoDB 的输入 input URI 连接属性：地址、端口、MyDB 数据库及 MyCollection 集合信息；设置 MongoDB 的输出 output URI 连接属性：地址、端口、test 数据库及 foo 集合信息；设置 MongoDB 的 input 输入类型为 MongoInputFormat；设置 MongoDB 的更新操作 mongo.input.update：数学小于 90 分就加 5 分。

2）调用 SparkContext 的 newAPIHadoopRDD 方法。将数学成绩低于 90 分的加 5 分以后的输出结果保存到 MongoDB test 数据库的 foo 集合。

3）对 newAPIHadoopRDD 查询 MongoDB 的结果遍历，打印输出。第一个元素是 Object ID，第二个元素是 MongoDB 中 MyDB 数据库 MyCollection 集合中数学成绩低于 90 分就加 5 分的记录。

```
/**
 * 把 Math 小于 90 的分数加 5 分
 * 修改器 $inc 可以对文档的数字型的键进行增减的操作。
 */
def updateMath(sc:SparkContext): Unit = {
    val mongoConfig = new Configuration()
mongoConfig.set("mongo.input.uri","mongodb://127.0.0.1:27017/MyDB.MyCollection")
mongoConfig.set("mongo.output.uri","mongodb://127.0.0.1:27017/test.foo")
mongoConfig.set("mongo.job.input.format","com.mongodb.hadoop.MongoInputFormat")
mongoConfig.set("mongo.input.update","{'Score.Math':{$lt:90}},{$inc:{'Score.Math':5}},{
upsert:false,multi:true}")  //内嵌文档
```

```scala
    val mongoRDD = sc.newAPIHadoopRDD(mongoConfig,classOf[MongoInputFormat],classOf[Object],classOf[BSONObject])
    mongoRDD.foreach(x => println(x._1 + " " + x._2))
}
```

操作结果如下图:

```
{ "_id" : ObjectId("5720e8dd6d110e7ec14b671a"), "ID" : "1", "Name" : "A", "Gender" : "F", "Birthday" : "1996-09-12" }
{ "_id" : ObjectId("5720e8f46d110e7ec14b671b"), "ID" : "2", "Name" : "B", "Gender" : "M", "Birthday" : "1995-12-23" }
{ "_id" : ObjectId("5720e8fb6d110e7ec14b671c"), "ID" : "3", "Name" : "C", "Gender" : "M", "Birthday" : "1996-10-29" }
{ "_id" : ObjectId("5720e9006d110e7ec14b671d"), "ID" : "4", "Name" : "D", "Gender" : "M", "Birthday" : "1995-02-25" }
{ "_id" : ObjectId("5720e9066d110e7ec14b671e"), "ID" : "5", "Name" : "E", "Gender" : "F", "Birthday" : "1997-06-06" }
{ "_id" : ObjectId("5720e9176d110e7ec14b671f"), "ID" : "24", "Name" : "X", "Info" : { "Gender" : "F", "Birthday" : "1997-06-06" }, "Score" : { "Math" : 90, "History" : 87 } }
{ "_id" : ObjectId("5720e9276d110e7ec14b6720"), "ID" : "25", "Name" : "Y", "Info" : { "Gender" : "M", "Birthday" : "1998-12-21" }, "Score" : { "Math" : 83, "History" : 96 } }
{ "_id" : ObjectId("5720e92e6d110e7ec14b6721"), "ID" : "26", "Name" : "Z", "Info" : { "Gender" : "M", "Birthday" : "1998-03-21" }, "Score" : { "Math" : 88, "Chinese" : 84 } }
```

(4) 示例四: 删除历史分数小于 90 分的键

1) 创建 Hadoop 的 Configuration 配置类, 设置 MongoDB 的输入 input URI 连接属性: 地址、端口、MyDB 数据库及 MyCollection 集合信息; 设置 MongoDB 的 input 输入类型为 MongoInputFormat; 设置 MongoDB 的更新操作 mongo.input.update: 删除历史分数小于 90 分的键值。

2) 调用 SparkContext 的 newAPIHadoopRDD 方法。

3) 对 newAPIHadoopRDD 查询 MongoDB 的结果遍历, 打印输出。第一个元素是 ObjectID, 第二个元素是删除历史分数小于 90 分的集合记录, 即 X 同学的历史分数是 87 分小于 90 分, 因此 X 同学记录中的历史分数键值对被删除。

```scala
/**
 * 删除 History 分数小于 90 的键值
 */
def removeHistory(sc:SparkContext): Unit = {
    val mongoConfig = new Configuration()
    mongoConfig.set("mongo.input.uri","mongodb://127.0.0.1:27017/MyDB.MyCollection")
    mongoConfig.set("mongo.job.input.format","com.mongodb.hadoop.MongoInputFormat")
    mongoConfig.set("mongo.input.update","{'Score.History':{$lt:90}},{$unset:{'Score.History':1}},{multi:true}") //内嵌文档

    val mongoRDD = sc.newAPIHadoopRDD(mongoConfig,classOf[MongoInputFormat],classOf[Object],classOf[BSONObject])
    mongoRDD.foreach(x => println(x._1 + " " + x._2))
}
```

操作结果如下:

```
{ "_id" : ObjectId("5720e8dd6d110e7ec14b671a"), "ID" : "1", "Name" : "A", "Gender" : "F", "Birthday" : "1996-09-12" }
{ "_id" : ObjectId("5720e8f46d110e7ec14b671b"), "ID" : "2", "Name" : "B", "Gender" : "M", "Birthday" : "1995-12-23" }
{ "_id" : ObjectId("5720e8fb6d110e7ec14b671c"), "ID" : "3", "Name" : "C", "Gender" : "M", "Birthday" : "1996-10-29" }
{ "_id" : ObjectId("5720e9006d110e7ec14b671d"), "ID" : "4", "Name" : "D", "Gender" : "M", "Birthday" : "1995-02-25" }
{ "_id" : ObjectId("5720e9066d110e7ec14b671e"), "ID" : "5", "Name" : "E", "Gender" : "F", "Birthday" : "1997-06-06" }
{ "_id" : ObjectId("5720e9176d110e7ec14b671f"), "ID" : "24", "Name" : "X", "Info" : { "Gender" : "F", "Birthday" : "1997-06-06" }, "Score" : { "Math" : 90 } }
{ "_id" : ObjectId("5720e9276d110e7ec14b6720"), "ID" : "25", "Name" : "Y", "Info" : { "Gender" : "M", "Birthday" : "1998-12-21" }, "Score" : { "Math" : 83, "History" : 96 } }
{ "_id" : ObjectId("5720e92e6d110e7ec14b6721"), "ID" : "26", "Name" : "Z", "Info" : { "Gender" : "M", "Birthday" : "1998-03-21" }, "Score" : { "Math" : 88, "Chinese" : 84 } }
```

第3章 Spark SQL操作多种数据源

（5）示例五：保存数据到mongodbMongoDB中

1）创建Hadoop的Configuration配置类，设置MongoDB的输入input URI连接属性：地址、端口、MyDB数据库及MyCollection集合信息；设置MongoDB的输出output URI连接属性：地址、端口、MyDB数据库及MyCollection集合信息；

2）调用Spark的parallelize方法生成data RDD。

3）遍历data，将data的数据写入到BasicBSONObject，BasicBSONObject对象obj放入name、age键值对，map方法遍历以后返回元组Key – Value（null, obj）键值对，元组的第一个元素为null，因为保存至Hadoop时第一个元素是NullWritable；第二个元素为BasicBSONObject对象obj。

4）调用SparkContext的saveAsNewAPIHadoopFile方法保存记录，saveAsNewAPIHadoopFile方法的第一个参数是MongoDB结果保存的路径，即保存在MyDB数据库的MyCollection集合中，第二个参数是输入Key的类型Any，第三个参数是输入Value的类型Any，第四个参数是输出的类型MongoOutputFormat[Any, Any]，第五个参数是Hadoop的配置类mongoConfig。

```
/**
  * 保存到mongodb中
  */
  def saveToMongo(sc:SparkContext): Unit = {
    val mongoConfig = new Configuration()
mongoConfig.set("mongo.input.uri","mongodb://127.0.0.1:27017/MyDB.MyCollection")
mongoConfig.set("mongo.output.uri","mongodb://127.0.0.1:27017/MyDB.MyCollection")

    val data = sc.parallelize(List(("Tom",31),("Jack",22),("Mary",25)))

    //使用MongoOutputFormat将数据写入到MongoDB中
    val rdd = data.map((elem) => {
      val obj = new BasicBSONObject()
      obj.put("name",elem._1.toString)
      obj.put("age",elem._2.toString)

      //转换后的结果,键值对,第一个是BSON的ObjectId,插入时可以指定为null,MongoDB
    //Driver在插入到MongoDB时,自动生成obj是BSONObject,是MongoDB Driver接收的插入对象
      (null,obj)
    })
    rdd.saveAsNewAPIHadoopFile("MyDB.MyCollection",classOf[Any],classOf[Any],classOf[MongoOutputFormat[Any,Any]],mongoConfig)
  }
```

保存的结果如下所示：

```
> db.MyCollection.find()
{ "_id" : ObjectId("5720e8dd6d110e7ec14b671a"), "ID" : "1", "Name" : "A", "Gender" : "F", "Birthday" : "1996-09-12" }
{ "_id" : ObjectId("5720e8f46d110e7ec14b671b"), "ID" : "2", "Name" : "B", "Gender" : "M", "Birthday" : "1995-12-23" }
{ "_id" : ObjectId("5720e8fb6d110e7ec14b671c"), "ID" : "3", "Name" : "C", "Gender" : "M", "Birthday" : "1996-10-29" }
{ "_id" : ObjectId("5720e9006d110e7ec14b671d"), "ID" : "4", "Name" : "D", "Gender" : "M", "Birthday" : "1995-02-25" }
{ "_id" : ObjectId("5720e9066d110e7ec14b671e"), "ID" : "5", "Name" : "E", "Gender" : "F", "Birthday" : "1997-06-06" }
{ "_id" : ObjectId("5720e9176d110e7ec14b671f"), "ID" : "24", "Name" : "X", "Info" : { "Gender" : "F", "Birthday" : "1997-06-06 " }, "Score" : { "Math" : 90, "History" : 87 } }
{ "_id" : ObjectId("5720e9276d110e7ec14b6720"), "ID" : "25", "Name" : "Y", "Info" : { "Gender" : "M", "Birthday" : "1998-12-21 " }, "Score" : { "Math" : 78, "History" : 96 } }
{ "_id" : ObjectId("5720e92e6d110e7ec14b6721"), "ID" : "26", "Name" : "Z", "Info" : { "Gender" : "M", "Birthday" : "1998-03-21 " }, "Score" : { "Math" : 83, "Chinese" : 84 } }
{ "_id" : ObjectId("5720f22456d4ae238c61444e"), "name" : "Jack", "age" : "22" }
{ "_id" : ObjectId("5720f22456d4ae238c61444d"), "name" : "Tom", "age" : "31" }
{ "_id" : ObjectId("5720f22456d4ae238c61444f"), "name" : "Mary", "age" : "25" }
```

Section 3.7 本章小结

本章详细阐述了 Spark SQL 对多种数据源的操作，Spark SQL 可加载 Parquet、Json、文本文件等各种数据源格式的数据；Spark SQL 可以操作各种数据库，例如：Spark SQL 操作 Hive、Spark SQL 操作 Json 数据集、Spark SQL 操作 HBase、SparkSQL 操作 MySQL、SparkSQL 操作 MongoDB 等等。通过本章的学习，读者可以熟练掌握 Spark SQL 对各种数据源的操作。

第 4 章　Parquet 列式存储

4.1　Parquet 概述

Spark SQL 目前已经在框架内部默认实现了对 JSON、HBase、Apache Parquet 等数据格式的支持，在 Spark SQL 读写数据的时候，用户可以手动指定数据的格式，例如指定读取数据文件来源是 JSON 格式的，而在输出的时候可以指定输出为 Apache Parquet 格式，从 JSON 到 Apache Parquet 不同格式的具体转换是 Spark SQL 框架自动完成的。接下来介绍 Apache Parquet，Apache Parquet 是面向分析型业务的列式存储格式，由 Twitter 和 Cloudera 合作开发，并于 2015 年 5 月成为 Apache 顶级项目。Parquet 是一种语言无关列式存储格式的文件类型，可以适配多种计算框架，而且不与任何一种数据处理框架绑定在一起，适配多种语言和组件。

4.1.1　Parquet 的基本概念

1. 列式存储

列式存储指数据是按列存储的，每一列数据单独存放，列式存储以流的方式在列中存储所有的数据，主要适合批量数据处理和即席查询。

下面看一个行式存储及列式存储的示意图，如图 4-1 所示。

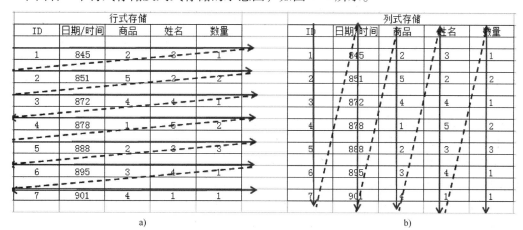

图 4-1　行式存储及列式存储示意图
a）行式存储　b）列式存储

图 4-1 中左侧的表格基于行式的存储，如表 4-1 所示。

表 4-1　行式存储

1	845	2	3	1	2	851	5	2	2	3	872	4	4	1	4	878	1	5	2

图 4-1 中右侧的表格基于列式的存储，如表 4-2 所示。

表 4-2　列式存储

1	2	3	4	…	845	851	872	878	…	2	5	4	1	…	3	2	4	5

从图 4-1 可以看出，基于行式的存储，依次存储每一行的数据，一张表的数据存放在一起。但列式存储的数据是按照列存储的，每一列单独存放，数据即是索引。如数据查询时只访问查询涉及的列，大大降低了系统 I/O，而且由于数据类型一致，方便压缩。

Parquet 是一种支持嵌套数据的列式存储格式。Parquet 元数据使用 Apache Thrift 进行编码。Parquet-format 项目包含创建 Parquet 文件的 readers 及 writers 所需的所有 Thrift 定义。Parquet 为 Hadoop 生态系统中的任何项目提供可压缩的列式数据表达、Parquet 使用 "record shredding and assembly algorithm"（基于 Dremel 的论文）算法来表示复杂的嵌套数据类型，通过列式压缩和编码技术降低了存储空间，提高了 I/O 效率。

2. Parquet 的 3 个核心组成部分

1）Storage Format（存储格式）：定义了 Parquet 内部的数据类型和存储格式。

2）Object Model Converters（对象模型转换器）：负责数据对象和数据类型之间的映射。这部分功能由 parquet-mr 项目来实现，主要完成外部对象模型与 Parquet 内部数据类型的映射，映射完成后 Parquet 会进行自己的 Column Encoding，然后存储 Parquet 格式文件。

3）Object Models（对象模型）：在 Parquet 中具有自己的 Object Model 定义的存储格式，例如，Avro 具有自己的 Object Model，但是 Parquet 在处理相关格式的数据时会使用自己的 Object Model 来完成具体数据的存储。

📖 Avro 是 Hadoop 中的一个子项目，是一个基于二进制数据传输高性能的中间件，是一个数据序列化的系统。Avro 可以将数据结构或对象转化成便于存储或传输的格式，适合远程或本地大规模数据的存储和交换。

3. 大数据分析技术栈中数据流水线处理的三种方式

业界对大数据分析技术栈的数据管道一般分为以下三种方式：

1）数据源→HDFS →MR/Hive/Spark（相当于 ETL）→HDFS Parquet→Spark SQL/Impala →ResultService（可以放在 DB 中，也有可能被通过 JDBC/ODBC 来作为数据服务使用）。

2）数据源→实时更新数据到 HBase/DB →导出为 Parquet 格式数据→Spark SQL/Impala →ResultService（可以放在 DB 中，也有可能被通过 JDBC/ODBC 来作为数据服务使用）。

上述第二种方式可以通过 Kafka + Spark Streaming + Spark SQL（强烈建议采用 Parquet 的方式来存储数据）方式取代。业界最期待的数据管道方式为下述第三种方式：

3）数据源→Kafka →Spark Streaming →Parquet →Spark SQL（ML、GraphX 等）→Parquet →其它各种 Data Mining 等。

由此可见，Parquet 格式数据在整个大数据分析过程中在数据管道中起着承上启下，数据格式转换的作用，处于数据管道的中间环节，其地位相当重要。

4.1.2　Parquet 数据列式存储格式应用举例

下面以通讯录地址本为例说明 Parquet 列式存储格式。

```
message AddressBook{
    required string owner;
    repeated string ownerPhoneNumbers;
    repeated group contacts{
        required string name;
        optional stringphoneNumber;
    }
}
```

整个通讯录地址本是一个嵌套结构，根结点是通讯录地址本（message），地址本里面包含多个字段［如业主 owner、业主电话 owner Phone Number、联系人（名字 Name、电话 PhoneNumber）］。每个字段包含 3 个属性：重复属性、字段类型、标识符。

其中重复属性可以是以下 3 种类型中的任意一种：

1）required（必需的，表示出现 1 次）。

2）optional（可选的，表示出现 0 次或者 1 次）。

3）repeated（重复的，表示出现 0 次或者多次）。

其中字段类型可以是以下两种：

1）group（组类型表示为一个组结构体，如联系人结构体中包括联系人的电话和名字）。

2）primitive（基本类型可以表示为 int、float、boolean、string 等）。

其中标识符可以解析为词法标识符单词：如业主名字、业主电话等。

在通讯录地址本嵌套结构中，每条记录表示一个人的通讯录，通讯录中必须登记业主；业主可以记录 0 个或多个电话号码；每个业主可以拥有 0 个或多个联系人，每个联系人必须记录名字，而联系人的电话号码可选。

通讯录地址本可以用图 4-2 来表示。

对通讯录地址本的树状结构图解释如下：

1）图 4-2 中，叶子结点分别为：业主 owner、业主电话 ownerPhoneNumbers、联系人名字 name、联系人电话 phoneNumber。

2）在逻辑上而言，模式（schema）实质上是一个表，如图 4-3 所示。

3）对于一个 Parquet 文件而言，数据会被分成 Row Group（里面包含很多 Column，每个 Column 就是这一列的数据），这样就构成了矩阵。

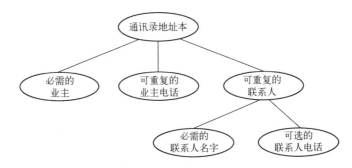

图 4-2 通讯录地址本的树状结构图

通讯录地址本			
业主	业主电话	联系人	
		联系人名字	联系人电话

图 4-3 schema 示意图

4) Column 具有几个非常重要的特性,例如:Repetition Level(重复级别)、Definition Level(定义级别)。Column 在 Parquet 中是以 Page(页)的方式存在的,Page 中有 Repetition Level、Definition Level 等内容。从根结点往叶子结点遍历的时候,会记录深度,这个就是 Definition Level。Definition Level 方便我们精准地找到数据。owner 是 required,所以将其 Definition Level 可以定义为 0。ownerPhoneNumber 结点没有叶子结点,定义为 1。Name 结点是 required,定义为 1。PhoneNumber 是 optional,有可能出现,也有可能不出现,定义为 2。Repetition Level 为重复级别。Owner 为 0,ownerPhoneNumber 为 1,name 和 phoneNumber 都是 1。

5) Row Group 在 Parquet 中是数据读写的缓存单元,所以对 Row Group 的设置会极大地影响 Parquet 的使用速度和效率。如果是分析日志的话,建议把 Row Group 的缓存大小配置成 256MB,而很多人的配置都是大于 1 GB,如果想最大化地提高运行效率,强烈建议 HDFS 的 Block 大小和 Row Group 一致。

6) 在实际存储的时候把一个树状结构,通过巧妙的编码算法,转换成二维表结构。Parquet 在存储时,会将 AddressBook 正向存储为 4 列,读取的时候会逆向还原出 AddressBook 对象,如表 4-3 所示。

表 4-3 二维表

列	最大定义层次	最大重复层次
业主	0(必需的)	0(不重复的)
业主电话	1	1
联系人.姓名	1(名字必需的)	1(联系人可重复)
联系人.电话	2(电话号可选)	1(联系人可重复)

7) Google 的 Dremel 系统解决了列式存储的问题,其核心思想是使用 "record shredding and assembly algorithm" 记录切碎和组装算法来表示复杂的嵌套数据类型,同时辅以按列的高效压缩和编码技术,实现降低存储空间。该算法就是把一行数据打碎,打碎之后按照自己的编码规则,再把数据组装起来。Parquet 就是基于 Dremel 系统的数据模型和算法来实现把

树状结构转换成图 4-3 所示的二维表结构。

8）Definition Level 是为了快速精准地找到数据，通过 Definition Level 可以精准地判断数据一定在哪个位置上。嵌套数据类型的特点是有些 field 可以是空的，也就是没有定义。如果一个 field 是定义的，那么它的所有父结点都是被定义的。从根结点开始遍历，当某一个 field 的路径上的结点开始是空的时候，记录当前的深度作为这个 field 的 Definition Level。如果一个 field 的 Definition Level 等于这个 field 的最大 Definition Level，就说明这个 field 是有数据的。对于 required 类型的 field 必须是有定义的，所以这个 Definition Level 是不需要的。在关系型数据中，optional 类型的 field 被编码成 0（表示空）和 1（表示非空，或者反之）。所以对于上个例子，Owner 是叶子结点，且是一定存储的，因此就将 Definition Level 定义为 0。因为 Contacts 为 repeated，所以 name 的 Definition Level 不可以为 0，只能为 1。phoneNumber 是 optional，因此它的 Definition Level 是 2。

9）什么是 Repetition Level？Repetition Level 用于记录该 field 的值是在哪一个深度上重复的。只有 repeated 类型的 field 需要 Repetition Level，optional 和 required 类型的不需要。Repetition Level = 0 表示开始一个新的 record。在关系型数据中，repetition level 总是 0。所以对于上个例子，Owner 为 0，ownerPhoneNumber 为 1，name 和 phoneNumber 都是 1。下面是更细致的表达：

```
Owner:
    Definition Level:0 叶子结点,所以定义为 0
    Repetition Level:0 不重复
ownerPhoneNumber:
    Definition Level:1 因为没有子结点了,所以定义为 1
    Repetition Level:1 重复
Name:
    Definition Level:1 此时不可以为 0,因为 Contacts 可能不存在
    Repetition Level:1
phoneNumber:optional
    Definition Level:2
    Repetition Level  1
```

通过上述 Definition Level、Repetition Level 的取值就可以精准地判断哪个数据一定在什么地方，以及映射成物理结构也比较容易，具体写到磁盘的时候就是 Definition Level、Repetion level 和 value 的构成。

例如：如果一条记录是如下的数据记录，包括两条 AddressBook，第一条 AddressBook 中有两个联系人，其中第一个联系人的姓名是 Spark，第二个联系人的电话是 18610086859；第二条 AddressBook 为空。

```
AddressBook {
    contacts: {
        name: "Spark"
```

```
            }
    contacts：{
            phoneNumber："18610086859"
            }
        }
AddressBook {
        }
```

分析上述记录，如图 4-4 中，R 表示重复级别，D 表示定义级别。对于 name："Spark"字段：其 R 重复级别是 0，表示是一个新的 record 记录，从根开始按照 schema 建立结构，D 定义级别是 1，name 是 required 的，因此不需要 definition level，所以沿用 contacts 的级别 1；对于 phoneNumber："18610086859" 字段：其 R 重复级别是 1，表示在第一级插入新值，即从第一条 AddressBook 的第一个 contacts 联系人之后插入一个新联系人记录，phoneNumber 是 optional 的，需要 definition level，这里 D 定义级别是 2；对于第二条 AddressBook，其 R 重复级别是 0，表示是一个新的 record 记录，D 定义级别也是 0，null 表示是空数据。

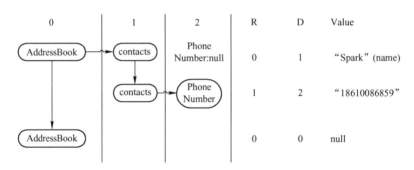

图 4-4 AddressBook 记录的序列化过程示意图

整理上述两条 AddressBook 的通讯信息如下图 4-5 所示。

Repetion level	Definition Level	Value
1	2	18610086859
0	1	"Spark"
0	0	NULL

图 4-5 Definition Level、Repetion level 和 Value 的构成表

Section 4.2 Parquet 的 Block 配置及数据分片

接下来讲解在 Parquet 中控制 Block 大小的配置参数，以及在 Spark SQL 中 Parquet 进行数据分片是如何实现的。通过在 Parquet 中控制 Block 大小可以设置更加适合的并行度，Parquet 的数据分片主要介绍 Parquet 数据分片的运行机制。

4.2.1　Parquet 的 Block 的配置

ParquetOutputFormat 是 Parquet 框架源代码中的类，ParquetOutputFormat 继承自 FileOutputFormat，在创建 ParquetOutputFormat 实例的时候，涉及 parquet.block.size 参数的配置。parquet.block.size 这个参数可以控制 Spark SQL 操作 Parquet 文件时候写入磁盘的大小数据。

如何对 parquet.block.size 进行配置？可以通过 SparkSQL 的 Conf 对象进行配置。配置方法为：

org.apache.spark.deploy.SparkHadoopUtil.get.conf.set("parquet.block.size","new value")

Parquet 的读写是以 Block 为单位的，所以通常建议将 Block 大小设置为 256 MB 或者 128 MB。当写入一个完整的 Block 时，由于在实现的时候做了非常大的 cache，需要把全部 cache 放入 Executor 中，这样 Parquet 就会非常耗内存。

Parquet 采用了非常大的压缩比例，在存储的时候对内存和磁盘空间的占用非常小，但是基于 Parquet 这种高度压缩的数据存储格式而言，每次把 Block 读进内存的时候实际数据是 parquet.block.size 设定值大小的好几倍。

4.2.2　Parquet 内部的数据分片

Spark 应用程序在 Spark 集群上运行，数据源是分布式的，Spark 在运行时在不同的结点需并行处理各个结点分区的数据。例如，在 Hadoop 文件中，Hadoop 文件的分区数 Split 的大小是由文件的总大小和分区数决定的；而在 Parquet 中，以 RowGroup 为基本单位，parquet.block.size 设置 parquet row group 大小，也就设置了分区的大小，Spark SqlNewHadoopPartition 获取的分区是从 Parquet 中得到的。同样的，Spark 从各种类型数据源（例如 Hadoop、Parquet、JDBC、RDD…）根据各自的分区算法获取到数据源的数据分区，数据分区具备数据本地性，数据存放在各个分布式结点上，Spark 应用程序就要在各个结点上并行处理不同的分区数据，一个分区对应一个执行任务，有多少个分区就有多少个任务并行度，进行分布式计算，并行处理数据更快。

从 Parquet 的角度考虑，在做分片的时候，Parquet 分片本身会变成 Partitions，在 Spark SQL 的源码中的 SqlNewHadoopPartition 类会继承 SparkPartition，在 SqlNewHadoopPartition 的 getPartitions 中的 inputFormatClass 实际上是 Parquet 包中的 ParquetInputFormat，getPartitions 中的 inputFormat 的 getSplits 是 ParquetInputFormat 的 getSplits，在 Spark SQL 中 Parquet 覆盖了 SparkPartition 的中 getSplits 的默认行为。

4.3　Parquet 序列化

本节介绍 Parquet 序列化的内容。

序列化是指把对象转换为二进制字节序列，反序列化是指把二进制字节序列恢复为对象。

例如，Spark 在 Task 执行及结果处理中执行具体 Task 的业务逻辑前会进行 4 次反序列化：

1）TaskDescription 的反序列化。
2）反序列化 Task 的依赖。
3）Task 的反序列化。
4）RDD 反序列化。

4.3.1　Spark 实施序列化的目的

Spark 之所以进行序列化，最重要的原因是内存空间有限（减少 GC 的压力，最大化地避免 Full GC 的产生，因为一旦产生 Full GC，则整个 Task 处于停止状态）、减少磁盘 I/O 的压力、减少网络 I/O 的压力。

什么时候会产生序列化和反序列化呢？发送磁盘 I/O 和网络通信的时候会序列化和反序列化，更为重要的需要考虑序列化和反序列化有另外两种情况：

a）Persist（Checkpoint）的时候必须考虑序列化和反序列化，例如，缓存到内存的时候只能使用 JVM 分配的 60% 的内存空间，此时好的序列化机制就至关重要了。

b）编程的时候。使用算子的函数操作，如果传入了外部数据，就必须执行序列化和反序列化。

4.3.2　Parquet 两种序列化方式

在 Spark SQL 中，Parquet 的序列化和反序列化有两种方式：

1）原生的 ParquetRecordReader 序列化和反序列化方式：Spark SQL 的 Parquet 序列化和反序列化非常消耗时间，在性能优化的时候，Parquet 序列化和反序列化一般要在 Spark SQL 读写文件的过程中消耗 60%～80% 的时间。

2）优化的 UnsafeRowParquetRecordReader 方式，优化的 RecordReader 类似钨丝计划，直接使用操作系统的内存，这样就省掉了序列化与反序列化本身。在 Spark 的 SqlNewHadoopRDD 类中，如果开启了 spark.sql.parquet.enableUnsafeRowRecordReader，Spark 在计算的时候，进行布尔值逻辑判断，如果数据源类型是 ParquetInputFormat，enableUnsafeRowParquetReader 是 true，且 tryInitialize 初始化成功，就创建一个优化的 RecordReader，即 UnsafeRowParquetRecordReader。UnsafeRowParquetRecordReader 中使用的 Row 为 UnsafeRow，UnsafeRow 直接操作原生内存，和钨丝计划一样，使用堆外内存，减少 GC 的消耗，减少 Java 对象对于数据包装冗余的存储。因为实际上在 Parquet 文件中具体的 flat 格式里面会有很多比较小的数据，所以不使用 Java Object 去封装的话，就可以节省很多的空间。在 UnsafeRow 中，直接操作 Platform 内存备份，这样就省略掉了序列化和反序列化。

原生的 ParquetRecordReader 跟 UnsafeRowParquetRecordReader 有什么区别呢？非常显著

的区别是批量反序列化，优化的 RecordReader 使用的是原生内存，批量反序列化，一次把 row group 数据读进来，这就极大地提高了性能。

Section 4.4 本章小结

本章讲解了 Parquet 的相关基本概念，对 Parquet 数据列式存储格式应用进行举例说明；讲解了 Parquet 的 Block 配置及数据分片，Parquet 序列化等内容。通过本章的学习，读者可以了解 Parquet 列式存储的相关原理及基础知识。

第5章　Spark SQL 内置函数与窗口函数

Spark SQL 是处理结构化数据的 Spark 模块，它提供了 DataFrame 这种编程的抽象，同时也可以作为分布式 SQL 查询引擎使用。而 DataFrame 是一种带有列名的分布式数据集合，我们可以把它理解为数据库中的一张表或者 R 语言、Python 语言中的 DataFrame，不过 Spark 在其底层做了众多优化，我们可以使用结构化的数据文件、数据库、数据仓库或者 RDD 来构造出 DataFrame。

在 Spark SQL 中，提供了大量的内置函数对数据进行分析。与 Spark SQL 的 API 不同的是，DataFrame 中的内置函数操作的返回结果是 Column 对象，这说明了 DataFrame 通过列来组织数据的分布式数据集（A distributed collection of data organized into named columns），这就为相对复杂的数据分析提供了极大地便利。例如，我们在操作 DataFrame 的方法中，可以随时调用内置函数进行业务数据的处理和分析，这对于我们构建极为复杂的业务逻辑算法极大地减少了时间成本，可以让开发者聚焦于数据分析上，对提高开发效率具有重大意义。

5.1 Spark SQL 内置函数

5.1.1 Spark SQL 内置函数概述

从 Spark 1.5.x 开始，Spark SQL 提供了大量的内置函数。使用时只需导入 org.apache.spark.sql.functions 即可。Spark SQL 内置函数包括聚合函数、集合函数、日期时间函数、数学函数、窗口函数、字符串函数、其他函数等各类函数。比如，在处理业务数据时，我们通常离不开 agg 聚合函数，使用 count、sum、avg、max、min 等操作，需要对数据的某些列进行分组，基于这些操作我们可以完成例如时间范围内商品销量统计、实时性气象指标统计甚至更为复杂的数据分析。在 DataFrame 的 API 中有许多重载的 agg 函数，下面是 Spark 1.6.0 下 DataFrame 类中 agg 函数的定义：

```
def agg(expr: Column, exprs: Column *): DataFrame = groupBy().agg(expr, exprs : _ *)
```

通过这段定义大致可以看出 DataFrame 的 agg 方法都需要先执行调用 groupBy 方法：

```
def groupBy(cols: Column *): GroupedData = {
  GroupedData(this, cols.map(_.expr), GroupedData.GroupByType)
}
```

groupBy 函数的返回值为 GroupedData，然后转过来再调用 GroupedData 类中的 agg 方法：

```
def agg(expr: Column, exprs: Column *): DataFrame = {
  toDF((expr + : exprs).map(_.expr))
}
```

第5章 Spark SQL内置函数与窗口函数

在 org. apache. spark. sql. functions 包下提供了大量的内置函数供开发者使用，总体上可以包含如下类型：

1）聚合函数，例如 countDistinct、sumDistinct 等。
2）集合函数，例如 sort_array、explode 等。
3）日期时间函数，例如 hour、quarter、next_day 等。
4）数学函数，例如 asin、atan、sqrt、tan、round 等。
5）窗口函数，例如 row_number、rank 等。
6）字符串函数，例如 concat、format_number、regexp_extract 等。
7）其他函数，例如 isNan、sha1、randn、callUDF 等。

表 5-1 列出了内置函数中最为常见的 agg 聚合函数，其他函数可以查阅源代码 org. apache. spark. sql. functions。

表 5-1 内置函数中最为常见的聚合函数

函数定义	使用说明
approxCountDistinct(e:Column, rsd:Double)	返回一个组中不同项目的近似去重数
approxCountDistinct(e:Column)	
avg(e:Column)	返回一个组中元素的平均值
collect_list(e:Column)	返回一组中所有元素的列表，包含重复元素
collect_set(e:Column)	返回一组中元素的列表，不包含重复值
corr(column1:Column, column2:Column)	返回两列的相关系数
count(e:Column)	返回一个组中元素的数量
countDistinct(expr:Column, exprs:Column*)	返回一个组中的不同元素的数量
first(e:Column)	返回一个组中第一个元素的值
kurtosis(e:Column)	返回一个组中的峰度
last(e:Column)	返回一个组中的最后一个元素值
max(e:Column)	返回一个组中元素的最大值
mean(e:Column)	返回一个组中元素的平均值
min(e:Column)	返回一个组中元素的最小值
skewness(e:Column)	返回一个组中元素的偏斜值
stddev(e:Column)	同 stddev_samp
stddev_pop(e:Column)	返回一个组中元素的总体标准差
stddev_samp(e:Column)	返回一个组中元素的总体标准差
sum(e:Column)	返回表达式中所有值的总和
sumDistinct(e:Column)	返回表达式中不同值的总和
var_pop(e:Column)	返回一个组中元素值的方差
var_samp(e:Column)	返回一个组中元素值的无偏方差
variance(e:Column)	同 var_samp

5.1.2 Spark SQL 内置函数应用实例

本节介绍两个 Spark SQL 内置函数应用案例：
- 商品订单交易查询案例，以 Scala 本地编程的方式实现。

- 电商交易项目综合案例，以 Spark Shell 编程方式实现。

1. 商品订单交易查询案例

下面我们通过 Scala 编程的方式，以一定时间范围内商品订单交易为例，在集成开发环境中熟悉 Spark 内置函数的简单使用。我们这里采用的是 Scala IDE，当然也可以采用 IntelliJ IDEA 等其他集成开发工具。

1）创建 Spark 配置对象 SparkConf，通过 setAppName 方法设置应用程序名称，setMaster 方法设置程序要链接的 Spark 集群的 Master URL，这里设置为 Local 模式，这就代表 Spark 程序在本地运行。

```
val conf = newSparkConf( )
conf.setAppName("SparkSQLInlineFunction")
conf.setMaster("local")
```

2）创建 SparkContext 对象，构建 Spark SQL 的上下文。

```
val sc = newSparkContext(conf)
val sqlContext = new SQLContext(sc)
```

这里需要注意的是，使用 Spark SQL 内置函数，就需要以 import sqlContext.implicits._ 的方式导入 SQLContext 下的隐式转换的内容。

3）这里我们手动创建一些数据来模拟一定时间范围内商品订单交易信息，以（交易日期，订单编号，商品编号，订单总额）为模型，在实际情况下会比模拟的数据复杂很多，最后通过 parallelize 的方式构建出订单 RDD 分布式集合对象。

```
val orderData = Array(
    "2016-03-27,000000001,00001230,1000",
    "2016-03-27,000000002,00001231,1600",
    "2016-03-27,000000003,00001230,900",
    "2016-03-28,000000004,00001231,19",
    "2016-03-21,000000005,00001237,450",
    "2016-03-28,000000006,00001237,3400",
    "2016-03-28,000000007,00001231,400",
    "2016-03-28,000000008,00001234,112",
    "2016-03-26,000000009,00001231,900",
    "2016-03-26,000000010,00001234,100",
    "2016-03-26,000000011,00001231,1001",
    "2016-03-28,000000012,00001236,2000",
    "2016-03-26,000000013,00001231,3500",
    "2016-03-29,000000014,00001237,250",
    "2016-03-28,000000015,00001231,120",
    "2016-03-29,000000016,00001238,240",
```

第5章 Spark SQL内置函数与窗口函数

```
            "2016-03-25,000000017,00001231,370",
            "2016-03-21,000000018,00001230,299"
        )
        val orderDataRDD = sc.parallelize(orderData)
```

4）对业务数据进行预处理生成 DataFrame，将创建的 RDD 转换为 DataFrame，这里把数组中的每个 String 类型用逗号切分生成 Row 类型。

```
        val orderDataRDDRow = orderDataRDD.map(
            row => {
                val splited = row.split(",")
                Row(splited(0),splited(1),splited(2),splited(3).toInt)
            }
        )
        val structTypes = StructType(Array(
            StructField("trade_date", StringType, true),
            StructField("order_no", StringType, true),
            StructField("product_no", StringType, true),
            StructField("amount", IntegerType, true)
        ))
        val orderDataDF = sqlContext.createDataFrame(orderDataRDDRow,structTypes)
```

5）使用 Spark SQL 提供的内置函数对 DataFrame 进行操作。需要注意的是，内置函数生成的是 Column 对象并且采用字节码生成技术（Bytecode Generaction，BG）的方式，在 Spark SQL 执行物理计划的时候对匹配的表达式采用特定的代码，动态编辑，然后运行。

这里使用的 Spark SQL 内置函数：
- countDistinct 函数：该函数是内置函数中的聚合函数，返回组中不同项的个数。
- agg 函数：这里通过指定一系列聚合列计算聚合。

然后基于构建的 DataFrame 对数据进行分析，我们按日期统计每天成交多少种商品：

```
        orderDataDF.groupBy("trade_date")
                   .agg('trade_date, countDistinct('product_no).as("product_number"))
        .show
```

运行代码并查看结果：

```
+----------+----------+--------------+
|trade_date|trade_date|product_number|
+----------+----------+--------------+
|2016-03-21|2016-03-21|             2|
|2016-03-25|2016-03-25|             1|
```

```
|2016-03-26|2016-03-26|              2|
|2016-03-27|2016-03-27|              2|
|2016-03-28|2016-03-28|              4|
|2016-03-29|2016-03-29|              2|
+----------+----------+---------------+
```

我们再统计一下每天订单成交总额：

```
orderDataDF.groupBy("trade_date")
           .agg('trade_date,sum('amount).as("sum_amount")).show
```

运行代码并查看结果：

```
+----------+----------+----------+
|trade_date|trade_date|sum_amount|
+----------+----------+----------+
|2016-03-21|2016-03-21|       749|
|2016-03-25|2016-03-25|       370|
|2016-03-26|2016-03-26|      5501|
|2016-03-27|2016-03-27|      3500|
|2016-03-28|2016-03-28|      6051|
|2016-03-29|2016-03-29|       490|
+----------+----------+----------+
```

注意：由于 Spark 的版本的不断升级，对于聚合函数参数的写法都有所不同，这里归纳出几种常见的写法：

```
orderDataDF.groupBy("trade_date").agg(min('amount))
orderDataDF.groupBy("trade_date").agg("amount"->"min")
orderDataDF.groupBy("trade_date").agg(Map("amount"->"min"))
orderDataDF.groupBy("trade_date").agg(min($"amount"))
```

上述写法中的内置函数包括：
- min 函数：内置函数中的聚合函数，返回组中表达式的最小值。
- agg 函数：这里通过指定列名的映射来计算聚合的方法。由此产生的 *DataFrame* 将包含分组列。可用的聚合方法是 avg，max，min，sum，count（"平均值""最大值""最小值""总数""计数"）。

2. 电商交易项目综合案例：
本案例进行电商交易项目综合查询应用：
1）在 Hive 中创建数据库及数据库表，在 tbDate 日期分类表、订单表 tbStock、订单明细表 tbStockDetail 中加载相应的数据记录。
2）使用 Spark SQL 基于 Hive 数据库表进行查询统计：查询订单表 tbStock 中的记录数，

第5章 Spark SQL内置函数与窗口函数

查询订单明细表 tbStockDetail 中的记录数。

3）在复杂统计中，我们使用内置函数 sum 函数来统计每年每个商品的销量总额并注册为临时表 T1。

4）在此基础上，我们使用内置函数 max 函数查询出每年的最大销售总额，并注册为临时表 T2。

5）基于上两步的临时表，通过年份（T1 表中的 theyear）、商品销量总额（T1 表中的 year_amount），T2 表中的最大销售总额（max_year_amount），关联查询出所有订单中每年的畅销商品。

1）除去过多的日志，通过 Apache 的 Log4J 设置日志消息的级别。

```
import org.apache.log4j.{Level,Logger}
Logger.getLogger("org.apache.spark").setLevel(Level.WARN)
Logger.getLogger("org.apache.spark.sql").setLevel(Level.WARN)
```

2）初始化 HiveContext，这里需要注意两点：当前 Spark 的版本是否支持 Hive，其次 Hive 的配置文件 hive-site.xml 已经存放到 Spark 主目录的 conf 目录下。

```
val hiveContext = new org.apache.spark.sql.hive.HiveContext(sc)
```

3）创建数据库和数据表。

在 Hive 中创建一个数据库 SALEDATA，用于存放数据表。

```
create database SALEDATA;

16/04/28 21:08:29 INFO parse.ParseDriver: Parsing command: create database SALEDATA
16/04/28 21:08:30 INFO parse.ParseDriver: Parse Completed
...
16/04/28 21:08:37 INFO log.PerfLogger: </PERFLOG method=Driver.run start=1461848912133
    end=1461848917561 duration=5428 from=org.apache.hadoop.hive.ql.Driver>
res0: org.apache.spark.sql.DataFrame = [result: string]
```

使用这个数据库 SALEDATA：

```
use SALEDATA;

16/04/28 21:09:23 INFO parse.ParseDriver: Parsing command: use SALEDATA
16/04/28 21:09:23 INFO parse.ParseDriver: Parse Completed
...
16/04/28 21:09:24 INFO log.PerfLogger: </PERFLOG method=Driver.run start=1461848963818
    end=1461848964117 duration=299 from=org.apache.hadoop.hive.ql.Driver>
res1: org.apache.spark.sql.DataFrame = [result: string]
```

创建表 tbDate，这个表定义了日期的分类：

```
create table tbDate (
dateID STRING,
theyearmonth STRING,
theyear STRING,
themonth STRING,
thedate STRING,
theweek STRING,
theweeks STRING,
thequot STRING,
thetenday STRING,
thehalfmonth STRING
) ROW FORMAT DELIMITED
FIELDS TERMINATED BY ','
LINES   TERMINATED BY '\n';
```

16/04/28 21:12:25 INFO parse.ParseDriver: Parsing command: create table tbDate (dateIDSTRING, theyearmonth STRING, theyear STRING, themonth STRING, thedate STRING, theweek STRING, theweeks STRING, thequot STRING, thetenday STRING, thehalfmonth STRING) ROW FORMAT DELIMITED FIELDS TERMINATED BY ','LINES   TERMINATED BY '

16/04/28 21:12:25 INFO parse.ParseDriver: Parse Completed

...

16/04/28 21:12:26 INFO ql.Driver: Starting command(queryId = root_20160428211225_5868f134 - d61d - 4217 - a584 - 6974a2ce1807): create table tbDate (dateID STRING, theyearmonth STRING, theyear STRING, themonth STRING, thedate STRING, theweek STRING, theweeks STRING, thequot STRING, thetenday STRING, thehalfmonth STRING) ROW FORMAT DELIMITED FIELDS TERMINATED BY ','LINES   TERMINATED BY '
'

...

16/04/28 21:12:26 INFO log.PerfLogger: </PERFLOG method = Driver.run start = 1461849145766 end = 1461849146808 duration = 1042 from = org.apache.hadoop.hive.ql.Driver>

res4: org.apache.spark.sql.DataFrame = [result: string]

创建表 tbStock，这个表定义了订单的信息：

```
create table tbStock (
ordernumber STRING,
locationID STRING,
dateID STRING
) ROW FORMAT DELIMITED
FIELDS TERMINATED BY ','
LINES   TERMINATED BY '\n';
```

第5章 Spark SQL内置函数与窗口函数

```
16/04/28 21:14:00 INFO parse.ParseDriver: Parsing command: create table tbStock (ordernumber
    STRING,locationID STRING,dateID STRING) ROW FORMAT DELIMITED FIELDS TERMI-
    NATED BY ',' LINES TERMINATED BY '
...
16/04/28 21:14:01 INFO log.PerfLogger: </PERFLOG method = Driver.run start = 1461849240847
    end = 1461849241040 duration = 193 from = org.apache.hadoop.hive.ql.Driver>
res5: org.apache.spark.sql.DataFrame = [result: string]
```

创建表 tbStockDetail，这个表定义了订单的明细：

```
create table tbStockDetail (
ordernumber STRING,
rownum INT,
itemID STRING,
qty INT,
price INT,
amount INT
) ROW FORMAT DELIMITED
FIELDS TERMINATED BY ','
LINES TERMINATED BY '\n';

16/04/28 21:15:33 INFO parse.ParseDriver: Parsing command: create table tbStockDetail (ordernum-
    ber STRING,rownum INT,itemID STRING,qty INT,price INT,amount INT) ROW FORMAT
    DELIMITED FIELDS TERMINATED BY ',' LINES TERMINATED BY '
'
16/04/28 21:15:33 INFO parse.ParseDriver: Parse Completed
...
16/04/28 21:15:34 INFO log.PerfLogger: </PERFLOG method = Driver.run start = 1461849333222
    end = 1461849334086 duration = 864 from = org.apache.hadoop.hive.ql.Driver>
res6: org.apache.spark.sql.DataFrame = [result: string]
```

4）分别加载 tbDate.txt、tbStock.txt、tbStockDetail.txt 这 3 个数据文件至 tbDate、tbStock、tbStockDetail 表中。

```
LOAD DATA LOCAL INPATH '/root/Downloads/data/ebusiness/tbDate.txt' OVERWRITE INTO TA-
    BLE tbDate;
LOAD DATA LOCAL INPATH '/root/Downloads/data/ebusiness/tbStock.txt' OVERWRITE INTO TA-
    BLE tbStock;
LOAD DATA LOCAL INPATH '/root/Downloads/data/ebusiness/tbStockDetail.txt' OVERWRITE IN-
    TO TABLE tbStockDetail;
```

注意：这里通过 OVERWRITE INTO 实现覆盖表原有的数据的目的。

5)查询统计。

在订单表 tbStock 中的记录数以及订单明细表 tbStockDetail 中的记录数的查询中没使用到 Spark SQL 内置函数,我们通过普通的 select 查询语句进行查询。

① 查询订单表 tbStock 中的记录数:

```
SELECT * FROM tbStock

scala > hiveContext.sql("SELECT * FROM tbStock").count;
16/04/28 21:22:18 INFO parse.ParseDriver: Parsing command: SELECT * FROM tbStock
16/04/28 21:22:18 INFO parse.ParseDriver: Parse Completed
16/04/28 21:22:20 INFO Configuration.deprecation: mapred.map.tasks is deprecated. Instead, use
    mapreduce.job.maps
16/04/28 21:22:22 INFO mapred.FileInputFormat: Total input paths to process : 1
res10: Long = 21154
```

② 查询订单明细表 tbStockDetail 中的记录数:

```
SELECT * FROM tbStockDetail

scala > hiveContext.sql("SELECT * FROM tbStockDetail").count;
16/04/28 21:23:56 INFO parse.ParseDriver: Parsing command: SELECT * FROM tbStockDetail
16/04/28 21:23:56 INFO parse.ParseDriver: Parse Completed
16/04/28 21:23:56 INFO mapred.FileInputFormat: Total input paths to process : 1
res11: Long = 287950
```

6)复杂统计。

本节中我们将使用内置函数 sum 函数来统计每年每个商品的销量总额;使用内置函数 max 函数查询出每年的最大销售总额。

① 查询出每年每个商品的销量总额。

每个订单可能对应不同的商品,所以需要通过订单表 tbStock 和订单明细表 tbStockDetail 中的 ordernumber 字段进行关联。

时间维度上需要根据年来统计销量总额,而 tbDate 定义了日期的分类信息,如年、月、日等,同时为保证后续按照时间维度进行查询,所以需要根据订单表 tbStock 中的 dateID 下单时间 ID 和 tbDate 进行关联,这样就需要关联 3 张表。

```
SELECT
    c.theyear, b.itemID,
    sum(b.amount) as year_amount
FROM
    tbStock a, tbStockDetail b, tbDate c
WHERE a.ordernumber = b.ordernumber AND
    a.dateID = c.dateID
GROUP BY c.theyear, b.itemID
```

第5章 Spark SQL内置函数与窗口函数

我们将这个 DataFrame 注册为临时表 T1 以供后续使用，其中 year_amount 为每年每个商品的销量总额。

```
scala > val DF1 = hiveContext.sql("SELECT c.theyear,b.itemID,sum(b.amount) as year_amount FROM
    tbStock a,tbStockDetail b,tbDate c WHERE a.ordernumber = b.ordernumber AND
    a.dateID = c.dateID GROUP BY c.theyear,b.itemID");
16/04/28 21:28:03 INFO parse.ParseDriver:Parsing command:SELECT c.theyear,b.itemID,
    sum(b.amount) as year_amount FROM tbStock a,tbStockDetail b,tbDate c WHERE
    a.ordernumber = b.ordernumber AND a.dateID = c.dateID GROUP BY c.theyear,b.itemID
16/04/28 21:28:03 INFO parse.ParseDriver:Parse Completed
DF1:org.apache.spark.sql.DataFrame = [theyear:string,itemID:string,year_amount:bigint]

scala > DF1.registerTempTable("T1");
scala > DF1.show
16/04/28 21:28:58 INFO mapred.FileInputFormat:Total input paths to process:1
16/04/28 21:28:58 INFO mapred.FileInputFormat:Total input paths to process:1
16/04/28 21:29:00 INFO mapred.FileInputFormat:Total input paths to process:1
+----------+---------------+-----------+
|   theyear|         itemID|year_amount|
+----------+---------------+-----------+
|      2006|  ZM5132963W0101|        698|
|      2006|  MD215300610101|      21102|
|      2006|  E2526248175402|       5162|
|      2006|  YA214339030101|       8291|
|      2006|  DF123303080209|        330|
|      2006|  WZ214336710201|        866|
|      2006|  DF124390290702|       1616|
|      2006|  ZX124385510101|        299|
|      2006|  04424449110101|       2134|
|      2006|  01326496590101|       5813|
|      2007|  YL324467200201|       2494|
|      2007|  G4325338953202|       1146|
|      2007|  YA214373080101|       8173|
|      2007|  BF127121990402|      12199|
|      2007|  YA214352210101|       5298|
|      2007|  YL524212313002|       1423|
|      2007|  CL126361160104|        696|
|      2007|  24125311990102|        594|
|      2007|  84526257690112|       2310|
|      2007|  77627236030106|      16987|
+----------+---------------+-----------+
only showing top 20 rows
```

② 查询出每年的最大销售总额。

在上一步查询的基础之上对 theyear 进行分组，运用 max 聚合计算出每年最大销售总额，代码如下：

```
SELECT
    T1.theyear,
    max(T1.year_amount) as max_year_amount
FROM T1
    GROUP BY T1.theyear
```

我们将这个 DataFrame 注册为临时表 T2 以供后续使用，其中 max_year_amount 为每年的最大销售总额，代码如下：

```
scala > val DF2 = hiveContext.sql("SELECT T1.theyear,max(T1.year_amount) as max_year_amount
    FROM T1 GROUP BY T1.theyear");
16/04/28 21:32:44 INFO parse.ParseDriver:Parsing command:SELECT T1.theyear,max(T1.year_
    amount) as max_year_amount FROM T1 GROUP BY T1.theyear
16/04/28 21:32:44 INFO parse.ParseDriver:Parse Completed
DF2:org.apache.spark.sql.DataFrame = [theyear:string,max_year_amount:bigint]
scala > DF2.registerTempTable("T2");
scala > DF2.show;
16/04/28 21:33:12 INFO mapred.FileInputFormat:Total input paths to process:1
16/04/28 21:33:12 INFO mapred.FileInputFormat:Total input paths to process:1
16/04/28 21:33:14 INFO mapred.FileInputFormat:Total input paths to process:1
+---------+----------------+
| theyear | max_year_amount|
+---------+----------------+
|    2004 |           53374|
|    2005 |           56569|
|    2006 |          113684|
|    2007 |           70226|
|    2008 |           97981|
|    2009 |           30029|
|    2010 |            4494|
+---------+----------------+
```

③ 查找出所有订单中每年的畅销商品。

基于上两步的临时表通过年份（T1 表中的 theyear）和统计出的商品销量总额（T1 表中的 year_amount 和 T2 表中的 max_year_amount）

将 T1 和 T2 进行关联，目的是为了查找商品的信息，代码如下：

```
SELECT
    T1.theyear,
```

第5章 Spark SQL内置函数与窗口函数

```
        T1. itemID,
        T2. max_year_amount
FROM
        T1, T2
WHERE   T1. theyear = T2. theyear AND
        T1. year_amount = T2. max_year_amount
ORDER BY T1. theyear
```

```
16/04/28 21:45:07 INFO parse. ParseDriver:Parsing command:SELECT T1. theyear,T1. itemID,
    T2. max_year_amount FROM T1,T2 WHERE   T1. theyear = T2. theyear AND
    T1. year_amount = T2. max_year_amount ORDER BY T1. theyear
16/04/28 21:45:07 INFO parse. ParseDriver:Parse Completed
16/04/28 21:45:11 INFO mapred. FileInputFormat:Total input paths to process:1
16/04/28 21:45:11 INFO mapred. FileInputFormat:Total input paths to process:1
16/04/28 21:45:11 INFO mapred. FileInputFormat:Total input paths to process:1
16/04/28 21:45:11 INFO mapred. FileInputFormat:Total input paths to process:1
16/04/28 21:45:18 INFO mapred. FileInputFormat:Total input paths to process:1
16/04/28 21:45:18 INFO mapred. FileInputFormat:Total input paths to process:1
+-------+------------------+----------------+
| theyear|           itemID| max_year_amount|
+-------+------------------+----------------+
|   2004|     JY424420810101|           53374|
|   2005|     24124118880102|           56569|
|   2006|     JY425468460101|          113684|
|   2007|     JY425468460101|           70226|
|   2008|     E2628204040101|           97981|
|   2009|     YL327439080102|           30029|
|   2010|     SQ429425090101|            4494|
+-------+------------------+----------------+
```

5.2 Spark SQL 窗口函数

5.2.1 Spark SQL 窗口函数概述

窗口函数，也可以称之为开窗函数或者分析函数。窗口函数（OVER 子句）为行定义一个窗口（将要操作的行的集合），它对一组值进行操作，不需要使用 GROUP BY 子句对数据进行分组，能够在同一行中同时返回基础行的列和聚合列。相对于窗口函数，SQL 聚合函数以 GROUP BY 查询对一组值进行聚合（例如 sum，avg，count 等操作），对数据进行分组后，对每个组查询只返回一行数据，不能同时返回基础列的数据，只能得到聚合列。

一般来说，传统的关系型数据库在聚合操作之后的行数都要小于聚合前的行数，对于窗口函数而言，操作前后的行数都是相等的。例如，我们要根据不同手机品牌下的不同型号进行分类并统计其个数，比如苹果手机有 5 种不同型号，华为手机有 4 种不同的型号，采用聚合函数统计时会产生两条记录，而采用窗口函数统计则会生成 9 条记录，即每一行都会产生窗口函数的结果。

在窗口函数出现之前存在着很多用 SQL 语句很难解决的问题，往往很多复杂的查询都要通过相关子查询或者复杂的存储过程来完成。直到后来一些传统意义上的关系型数据库引入了窗口函数，使得这些复杂的查询变得尤为容易。所谓窗口就是指用户指定的一组行。窗口函数计算从窗口派生的结果集中各行的值，分别应用于每个分区，并为每个分区重新启动计算。

Spark SQL 也为我们提供了大量的窗口函数，如表 5-2 所示：

表 5-2 常用的窗口函数

函数名称	函数功能
cume_dist()	返回窗口分区内的累积值分布，例如在当前行下的行分片
dense_rank()	返回窗口分区中的行的排名，排名无间隔。密集排名和排名的区别：例如，第一名有 1 人，第二名并列有 3 人，那么在密集排名中下一个人的排名是第三名；而在一般排名中，下一个人是第五名
lag(Column e, int offset)	返回当前行'offset'之前的行的值，如果当前行之前的行少于 offset 行，返回 null 值
lag(Column e, int offset, Object defaultValue)	返回当前行'offset'之前的行的值，如果当前行之前的行少于 offset 行，返回 null 值
lag(String columnName, int offset)	返回当前行'offset'之前的行的值，如果当前行之前的行少于 offset 行，返回 null 值
lag(String columnName, int offset, Object defaultValue)	返回当前行'offset'之前的行的值，如果当前行之前的行少于 offset 行，返回 null 值
lead(Column e, int offset)	返回当前行'offset'之后的行的值，如果当前行之后的行少于 offset 行，返回 null 值
lead(Column e, int offset, ObjectdefaultValue)	返回当前行'offset'之后的行的值，如果当前行之后的行少于 offset 行，返回 null 值
lead(String columnName, int offset)	返回当前行'offset'之后的行的值，如果当前行之后的行少于 offset 行，返回 null 值
lead(String columnName, int offset, Object defaultValue)	返回当前行'offset'之后的行的值，如果当前行之后的行少于 offset 行，返回 null 值
ntile(int n)	在一个有序的窗口分区返回 ntile 组 ID（从 1 到'n'）。例如，如果 n 是 4，第一部分的行将得到值 1，第二部分行将获得值 2，第三部分将获得值 3，而最后一部分获得值 4
percent_rank()	返回相对排名等同于 SQL 的 percent_rank 功能
rank()	返回窗口分区中的行的排名
row_number()	返回在窗口分区中从 1 开始的序列号

Spark SQL 窗口函数的语法如下：

```
functionName() OVER (PARTITION BY column1,column2,…columnN ORDER BY
    field1,field2,field3) AS alias;
```

OVER 关键字表示把函数当成窗口函数而不是聚合函数。对于查询结果的每一行都返回

第5章　Spark SQL内置函数与窗口函数

所有符合条件的行的条数。OVER 关键字后的括号中还经常添加选项，用于改变进行聚合运算的窗口范围。如果 OVER 关键字后的括号中的选项为空，则窗口函数会对结果集中的所有行进行聚合运算。

从语法层面上来看，Spark SQL 窗口函数和传统的关系型数据下的窗口函数的写法基本一致。

5.2.2　Spark SQL 窗口函数分数查询统计案例

本节数学考试分数查询案例中，将综合应用 Spark SQL 的窗口函数，例如：row_number()、rank()、percent_rank()、tile、cume_dist、lag、lead 等函数，统计班级内每位同学考试成绩的名次、百分比排名及相关统计功能：

1）将考试分数文件上传到 HDFS 系统（考试分数文件数据信息如表 5-3 所示）。

表 5-3　考试分数数据信息

班级	classNo
学生姓名	studentName
成绩	score

2）启动 Hive metastore 服务，启动 Spark shell 。

3）创建 hiveContext，创建数据库表 math_score，将 HDFS 分数文件加载到 Hive 数据库表中。

4）使用窗口函数查询班级内每位同学的考试排名情况。

1. 将考试分数文件上传到 HDFS 系统并将 HDFS 分数文件加载到 Hive 数据库表

1）在 HDFS 中创建文件夹，代码如下。

```
root@ Master:/usr/local/hive/apache-hive-1.2.1-bin/bin# hadoop dfs -mkdir -p
    /library/SparkSQL/function/mathscore
DEPRECATED:Use of this script to executehdfs command is deprecated.
Instead use thehdfs command for it.
16/04/23 17:35:00 WARN util.NativeCodeLoader:Unable to load native-hadoop library for your
    platform... using builtin-java classes where applicable
```

2）将数据文件 math_score.txt 上传到 HDFS 中，代码如下。

```
root@ Master:/usr/local/hive/apache-hive-1.2.1-bin/bin# hadoop dfs -put
    /root/Downloads/data/math_score.txt/library/SparkSQL/function/mathscore
DEPRECATED:Use of this script to executehdfs command is deprecated.
Instead use thehdfs command for it.
16/04/23 17:37:40 WARN util.NativeCodeLoader:Unable to load native-hadoop library for your
    platform... using builtin-java classes where applicable
```

3）启动 Hive 的 metastore 服务，代码如下。

root@ Master:/usr/local/hive/apache - hive - 1.2.1 - bin/bin# hive - - service metastore > metastore. log 2 >& 1& [1] 4518

4）进入 Spark 的 sbin 下启动 Spark 集群，代码如下。

root@ Master:/usr/local/spark/spark - 1.6.3 - bin - hadoop2.6/sbin# ./start - all.sh
starting org. apache. spark. deploy. master. Master,logging to /usr/local/spark/spark - 1.6.3 - bin -
　　hadoop2.6/logs/spark - root - org. apache. spark. deploy. master. Master - 1 - Master. out
Worker2:starting org. apache. spark. deploy. worker. Worker,logging to /usr/local/spark/spark - 1.6.3 -
　　bin - hadoop2.6/logs/spark - root - org. apache. spark. deploy. worker. Worker - 1 - Worker2. out
Worker1:starting org. apache. spark. deploy. worker. Worker,logging to /usr/local/spark/spark - 1.6.3 -
　　bin - hadoop2.6/logs/spark - root - org. apache. spark. deploy. worker. Worker - 1 - Worker1. out

5）进入 Spark 的 bin 目录下，采用 ./spark - shell - - master spark://Master:7077 启动 Spark - Shell。

6）创建 hiveContext 上下文。

scala > val hiveContext = new org. apache. spark. sql. hive. HiveContext(sc)
16/04/23 17:47:58 INFO hive. HiveContext:Initializing execution hive,version 1.2.1
16/04/23 17:47:58 INFO client. ClientWrapper:Inspected Hadoop version:2.6.0
16/04/23 17:47:58 INFO client. ClientWrapper:Loaded org. apache. hadoop. hive. shims.
　　Hadoop23Shims for Hadoop version 2.6.0
16/04/23 17:47:59 INFO hive. metastore:Mestastore configuration hive. metastore.
　　warehouse. dir changed from file:/tmp/spark - ada2fb30 - 77ad - 415a - a2d9 - a29f1fe562be/
　　metastore to file:/tmp/spark - c01e398a - 4bc4 - 4411 - ab4d - fb7c5679459c/metastore
16/04/23 17:47:59 INFO hive. metastore:Mestastore configuration javax. jdo. option. ConnectionURL
　　changed from jdbc:derby:;databaseName = /tmp/spark - ada2fb30 - 77ad - 415a - a2d9 -
　　a29f1fe562be/metastore;create = true to jdbc:derby:;databaseName = /tmp/spark - c01e398a -
　　4bc4 - 4411 - ab4d - fb7c5679459c/metastore;create = true
16/04/23 17:47:59 INFO metastore. HiveMetaStore:0:Shutting down the object store...
...
hiveContext:org. apache. spark. sql. hive. HiveContext = org. apache. spark. sql. hive. HiveContext@ 3683c47f

7）创建数据表，代码如下：

scala > hiveContext. sql("create table math_score(classNo STRING,studentName STRING,score INT)
　　ROW FORMAT DELIMITED FIELDS TERMINATED BY ','LINES TERMINATED BY '\n' ")
16/04/23 17:51:21 INFO parse. ParseDriver:Parsing command:create table math_score(classNo
　　STRING,studentName STRING,score INT)ROW FORMAT DELIMITED FIELDS
　　TERMINATED BY ','LINES TERMINATED BY '
'
16/04/23 17:51:21 INFO parse. ParseDriver:Parse Completed
...
16/04/23 17:51:26 INFO log. PerfLogger:</PERFLOG method = Driver. run start = 1461405081957

第5章 Spark SQL内置函数与窗口函数

```
        end = 1461405086780 duration = 4823 from = org. apache. hadoop. hive. ql. Driver >
res6:org. apache. spark. sql. DataFrame = [result:string]
```

8）加载数据，代码如下：

```
scala > hiveContext. sql ( " LOAD  DATA  INPATH  '/library/SparkSQL/function/mathscore/math_
score. txt '
    INTO TABLE math_score" );
16/04/23 18:10:17 INFO parse. ParseDriver:Parsing command:LOAD DATA INPATH
    '/library/SparkSQL/function/mathscore/math_score. txt 'INTO TABLE math_score
16/04/23 18:10:17 INFO parse. ParseDriver:Parse Completed
...
16/04/23 18:10:21 INFO log. PerfLogger: </PERFLOG method = Driver. run start = 1461406217913
        end = 1461406221416 duration = 3503 from = org. apache. hadoop. hive. ql. Driver >
res4:org. apache. spark. sql. DataFrame = [result:string]
```

查看数据结果如下：

```
+-----------+------------+------+
|    classNo|  studentName| score|
+-----------+------------+------+
|  DTSX0001 |        Jack |  100 |
|  DTSX0001 |       Smith |   95 |
|  DTSX0001 |        Lady |   95 |
|  DTSX0001 |       Curry |   90 |
|  DTSX0002 |        John |   92 |
|  DTSX0002 |       Peter |   92 |
|  DTSX0002 |        Alex |   85 |
|  DTSX0002 |      Lowery |   76 |
|  DTSX0002 |       James |   65 |
+-----------+------------+------+
```

2. 使用窗口函数查询班级内每位同学的考试排名情况

在本节考试分数查询统计中，我们将使用 Spark SQL 的窗口函数 row_number( )统计班级内每位同学考试成绩的名次；分别使用窗口函数 rank( )、dense_rank 统计学生的考试名次；使用窗口函数 percent_rank( )进行百分比排名；使用窗口函数 ntile 函数将分组数据按照顺序切分成 n 片；使用窗口函数 cume_dist 函数计算小于等于当前值的行数占分组内总行数的比例；使用窗口函数 lag 函数用于统计窗口内往上第 n 行的值；使用窗口函数 lead 函数统计窗口内往下第 n 行的值。

（1）def row_number( ):Column

该函数会从 1 开始按照由小到大的顺序生成分组内记录的序列。比如，我们按照成绩降序排列，生成每个班级内每位同学成绩的名次，代码如下：

```
SELECT
classNo,studentName,score,
    row_number()OVER(PARTITION BYclassNo ORDER BY score DESC)AS rn
FROM math_score
```

执行结果如下：

classNo	studentName	score	rn
DTSX0001	Jack	100	1
DTSX0001	Smith	95	2
DTSX0001	Lady	95	3
DTSX0001	Curry	90	4
DTSX0002	John	92	1
DTSX0002	Peter	92	2
DTSX0002	Alex	85	3
DTSX0002	Lowery	76	4
DTSX0002	James	65	5

ROW_NUMBER()的应用场景非常多，通常可以用于 Top N 排名的统计，例如网易云音乐按照不同的音乐类别统计出榜单的 TOP 10。在这里需要找出每个班级中成绩排名前三名的学生，代码如下：

```
SELECT tmp.*(
SELECT classNo,studentName,score,
row_number()OVER(PARTITION BYclassNo ORDER BY score DESC)AS rn
FROM math_score)tmp WHERE tmp<4
```

(2) def rank():Column 和 def dense_rank():Column

row_number 虽然能进行排名，但是在排序中出现相同值（例如并列第一名）的情况下仍依次排序，我们可能会分别采用 rank 和 dense_rank 进行排名，而这会对整个排名会产生一定的影响。rank 函数在遇到排名相等的情况下会在名次中留下空位；而 dense_rank 在排名相等时在名次中不会留下空位。

我们分别用3个窗口函数 row_number()、rank()、dense_rank()函数对考试分数进行排名统计。

row_number()、rank()、dense_rank()窗口函数分别查询统计如下：

```
SELECT
classNo,studentName,score,
    row_number()OVER(PARTITION BYclassNo ORDER BY score DESC)AS rn1,
```

```
        rank() OVER(PARTITION BY classNo ORDER BY score DESC) AS rn2,
        dense_rank() OVER(PARTITION BY classNo ORDER BY score DESC) AS rn3
    FROM math_score
```

执行结果如下：

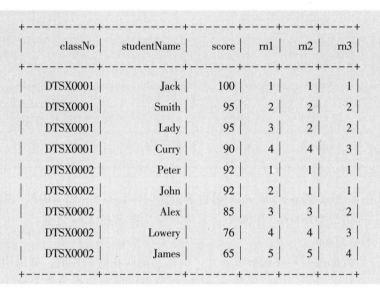

- rn1 列是 row_number 函数作用的结果。
- rn2 列是 rank 函数作用的结果；班级号为 DTSX0001 的班级中，Smith 和 Lady 同为 95 分，并列排名第 2，由于前两名有 3 人，会在第 3 名留下空位，所以 Curry 排名第 4，后续的排名按照这个规则进行排名。
- rn3 列是 dense_rank 函数作用的结果；班级号为 DTSX0002 的班级中，Peter 和 John 同为 92 分，并列第一，而后续的排名不会留下空位，即使有多人并列排名，Alex 在有两个并列第一的情况下仍然排名第 2。

（3）def percent_rank():Column

percent_rank 函数是在分组内中计算：当前记录的 rank 值 −1/分组内总行数 −1。rank_ 列计算使用窗口函数 rank() 计算班级中每位同学的排名；count_ 列计算每个班级的同学总人数；percent_rank() 列计算当前记录的 rank 值 −1/分组内总行数 −1 的百分比排名。

```
        SELECT
        classNo,studentName,score,
            rank() OVER(PARTITION BY classNo ORDER BY score DESC) AS rank_,
            count(1) OVER(PARTITION BY classNo) AS count_,
            percent_rank() OVER(PARTITION BY classNo ORDER BY score DESC) AS rn
        FROM math_score
```

执行结果如下：

```
+---------+-------------+-------+------+-------+-------------------+
| classNo | studentName | score | rank_| count_|                 rn|
+---------+-------------+-------+------+-------+-------------------+
|DTSX0001 |         Jack|    100|     1|      4|                0.0|
|DTSX0001 |        Smith|     95|     2|      4| 0.3333333333333333|
|DTSX0001 |         Lady|     95|     2|      4| 0.3333333333333333|
|DTSX0001 |        Curry|     90|     4|      4|                1.0|
|DTSX0002 |         John|     92|     1|      5|                0.0|
|DTSX0002 |        Peter|     92|     1|      5|                0.0|
|DTSX0002 |         Alex|     85|     3|      5|                0.5|
|DTSX0002 |       Lowery|     76|     4|      5|               0.75|
|DTSX0002 |        James|     65|     5|      5|                1.0|
+---------+-------------+-------+------+-------+-------------------+
```

例如，DTSX0002 中的 Alex 的 percent_rank 函数下值为 $(3-1)/(5-1)=0.5$，说明 Alex 考试分数的排名情况，即班级 DTSX002 有 50%（一半）学生排名在 Alex 之前。

(4) def ntile(n:Int):Column

ntile 函数用于将分组数据按照顺序切分成 n 片，返回当前所在的切片值，如果切片不均匀，则默认增加第一个切片的分布。窗口函数 ntile(2)将分组内所有数据分成 2 片，以及按 classNo 分区，返回当前所在的切片值作为 rn1 列；窗口函数 ntile(3)将分组内所有数据分成 3 片，以及按 classNo 分区，返回当前所在的切片值作为 rn2 列；窗口函数 ntile(3)将分组内所有数据分成 3 片，返回当前所在的切片值作为 rn3 列；

```
SELECT
classNo,studentName,score,
ntile(2)OVER(PARTITION BY classNo ORDER BY score DESC)AS rn1,
ntile(3)OVER(PARTITION BY classNo ORDER BY score DESC)AS rn2,
ntile(3)OVER(ORDER BY score DESC)AS rn3
FROM math_score
```

执行结果如下：

```
+---------+-------------+-----+---+---+---+
| classNo | studentName |score|rn1|rn2|rn3|
+---------+-------------+-----+---+---+---+
|DTSX0001 |         Jack|  100|  1|  1|  1|
|DTSX0001 |        Smith|   95|  1|  1|  1|
|DTSX0001 |         Lady|   95|  2|  2|  1|
|DTSX0002 |         John|   92|  1|  1|  2|
|DTSX0002 |        Peter|   92|  1|  1|  2|
```

```
|   DTSX0001  |     Curry     | 90 | 2 | 3 | 2 |
|   DTSX0002  |     Alex      | 85 | 1 | 2 | 3 |
|   DTSX0002  |     Lowery    | 76 | 2 | 2 | 3 |
|   DTSX0002  |     James     | 65 | 2 | 3 | 3 |
+-------------+---------------+----+---+---+---+
```

- rn1 是根据班级号分组的，然后按照成绩由高到低排序后切片成 2 份。
- rn2 是根据班级号分组的，然后按照成绩由高到低排序后切片成 3 份。
- rn3 把所有记录作为一个组，然后按照成绩由高到低排序后切片成 3 份。

以列 rn1 为例，即以班级号为 DTSX0001 这一组为例，该组一共 4 条记录被切分成 2 份，这样每个分片有两条记录，所以在该组中，Jack 和 Smith 的成绩位于第一个分片，值为 1，Lady 和 Curry 的成绩位于第二个分片，所以值为 2，这是均匀切片的情况。

我们来看不均匀切片的情况，在班级号为 DTSX0002 这一组中总计 5 条记录被切片成 2 份，5 除以 2 余 1，这一条记录加入在第一个分片中，就变了第 1 个分片有 3 条记录，第 2 个分片有 2 条记录，所以 John、Peter、Alex 的成绩位于第 1 个分片中，分片值为 1，Lowery 和 James 位于第 2 个分片中，所以分片值为 2。

5. def cume_dist( ):Column

cume_dist 函数用于计算小于等于当前值的行数占分组内总行数的比例，这在实际业务场景中非常常见，比如，统计小于等于当前薪水的人数所占总人数的比例。使用窗口函数 cume_dist( )统计考试分数在 Lowery 同学分数之前的人数在总人数的占比，作为 rn1 列；使用窗口函数 cume_dist( ) 及使用 PARTITION BY 按班级分组，统计考试分数在 Lowery 同学分数之前的人数在班级分组人数的占比，作为 rn2 列。

```
SELECT
classNo,studentName,score,
cume_dist( )OVER( ORDER BY score DESC)AS rn1,
cume_dist( )OVER( PARTITION BY classNo ORDER BY score DESC)AS rn2
FROM math_score
```

执行结果如下：

```
+-----------+-------------+------+--------------------+------+
|  classNo  | studentName |score |         rn1        | rn2  |
+-----------+-------------+------+--------------------+------+
|  DTSX0001 |    Jack     | 100  | 0.1111111111111111 | 0.25 |
|  DTSX0001 |    Smith    |  95  | 0.3333333333333333 | 0.75 |
|  DTSX0001 |    Lady     |  95  | 0.3333333333333333 | 0.75 |
|  DTSX0001 |    Curry    |  90  | 0.6666666666666666 | 1.0  |
|  DTSX0002 |    Peter    |  92  | 0.5555555555555556 | 0.4  |
|  DTSX0002 |    John     |  92  | 0.5555555555555556 | 0.4  |
```

```
|          |    DTSX0002    |    Alex    | 85 | 0.7777777777777778 | 0.6 |
|          |    DTSX0002    |    Lowery  | 76 | 0.8888888888888888 | 0.8 |
|          |    DTSX0002    |    James   | 65 |                1.0 | 1.0 |
+----------+----------------+------------+----+--------------------+-----+
```

- 列 rn1 是把当前所有记录作为一组的，按成绩由高到低来计算 cume_dist 的值。
- 列 rn2 是根据班级号分组的，按成绩由高到低来计算 cume_dist 的值，例如班级号为 DTSX0002 这一组中分数按降序排列，对于 Lowery 而言，排在 76 分前之前的记录数有 4 条，DTSX0002 的总记录条数为 5 条，所以该行的 cume_dist 值为 4/5 = 0.8。

(6) def lag(e:Column,offset:Int,defaultValue:Any):Column

lag 函数用于统计窗口内往上第 n 行值；第一个参数为列名，第二个参数为往上第 n 行（可选，默认为 1），第三个参数为默认值（当往上第 n 行为 NULL 时候，取默认值，如不指定，则为 NULL）。使用窗口函数 row_number() 计算班级中每位同学考试分数的排名，作为 rownum 列；使用窗口函数 lag(score,1) 获取班级中分数在此同学分数之前 1 位的同学的分数，作为 rn1；使用窗口函数 lag(score,2) 获取班级中分数在此同学分数之前 2 位的同学的分数，作为 rn2。

```
SELECT
classNo,studentName,score
    row_number()OVER(PARTITION BYclassNo ORDER BY score DESC)AS rownum,
    lag(score ,1,0)OVER(PARTITION BYclassNo ORDER BY score DESC)AS rn1,
    lag(score ,2)OVER(PARTITION BYclassNo ORDER BY score DESC)AS rn2
FROM math_score
```

执行结果如下：

```
+----------+-------------+-------+--------+------+------+
|  classNo | studentName | score | rownum |  rn1 |  rn2 |
+----------+-------------+-------+--------+------+------+
| DTSX0001 |     Jack    |  100  |    1   |   0  | null |
| DTSX0001 |     Smith   |   95  |    2   | 100  | null |
| DTSX0001 |     Lady    |   95  |    3   |  95  | 100  |
| DTSX0001 |     Curry   |   90  |    4   |  95  |  95  |
| DTSX0002 |     Peter   |   92  |    1   |   0  | null |
| DTSX0002 |     John    |   92  |    2   |  92  | null |
| DTSX0002 |     Alex    |   85  |    3   |  92  |  92  |
| DTSX0002 |     Lowery  |   76  |    4   |  85  |  92  |
| DTSX0002 |     James   |   65  |    5   |  76  |  85  |
+----------+-------------+-------+--------+------+------+
```

- 列 rn1 是根据班级进行分组的，按成绩由高到低来计算 lag 函数的值，例如在班级号为 DTSX0001 这个组中对于成绩最高的 Jack 来说，往上 1 行不存在，所以此时的值为我们设置的默认值为 0，对于 Smith 来说往上 1 行为 Jack 的记录，所以 lag 的值为 100。

- 列 rn2 是往上查找两条记录，比如在班级号为 DTSX0002 的这一组中，对于 John 来说，往上两行不存在，且没有设置默认值，此时为 NULL，对于 James 来说，往上查找两行为 Alex 的记录，所以 lag 值为 85。

(7) def lead(e:Column,offset:Int,defaultValue:Any):Column

lead 与 lag 相反，该函数用于统计窗口内往下第 n 行值：第一个参数为列名，第二个参数为往下第 n 行（可选，默认为 1），第三个参数为默认值（当往下第 n 行为 NULL 时候，取默认值，如果不指定，则为 NULL）。使用窗口函数 row_number() 计算班级中每位同学考试分数的排名，作为 rownum 列；使用窗口函数 lead(score,1) 获取班级中分数在当前的同学分数之后 1 位的同学的分数，作为 rn1；使用窗口函数 lead(score,2) 获取班级中分数在当前的同学分数之后 2 位的同学的分数，作为 rn2。

```
SELECT
classNo,studentName,score,
    row_number() OVER(PARTITION BY classNo ORDER BY score DESC) AS rownum,
    lead(score ,1,0) OVER(PARTITION BY classNo ORDER BY score DESC) AS rn1,
    lead(score ,2) OVER(PARTITION BY classNo ORDER BY score DESC) AS rn2
FROM math_score
```

执行结果如下：

```
+---------+-------------+-------+--------+-----+-----+
| classNo | studentName | score | rownum | rn1 | rn2 |
+---------+-------------+-------+--------+-----+-----+
| DTSX0001|        Jack |   100 |      1 |  95 |  95 |
| DTSX0001|       Smith |    95 |      2 |  95 |  90 |
| DTSX0001|        Lady |    95 |      3 |  90 | null|
| DTSX0001|       Curry |    90 |      4 |   0 | null|
| DTSX0002|       Peter |    92 |      1 |  92 |  85 |
| DTSX0002|        John |    92 |      2 |  85 |  76 |
| DTSX0002|        Alex |    85 |      3 |  76 |  65 |
| DTSX0002|      Lowery |    76 |      4 |  65 | null|
| DTSX0002|       James |    65 |      5 |   0 | null|
+---------+-------------+-------+--------+-----+-----+
```

列 rn1 是根据班级进行分组，获取班级中分数在当前的同学分数之后 1 位的同学的分数，例如在班级号为 DTSX0001 这个组中对于成绩最高的 Jack 来说，往下 1 行为 Smith 的记录，所以 lead 的值为 95；对于 Curry 来说往下 1 行不存在且有设置默认值，此时为 0。

列 rn2 是往下查找 2 条记录，比如在班级号为 DTSX0002 的这一组中，对于 Lowery 来说往下 2 行不存在且没有设置默认值,此时为 NULL,对于 Alex 来说往下查找 2 行为 James 的记录,所以 lead 值为 65。

5.2.3 Spark SQL 窗口函数 NBA 常规赛数据统计案例

下面我们以 2014~2015 和 2015~2016 赛季 NBA 常规赛为例,对窗口函数进行数据统计的应用实践。

1. NBA 常规赛数据说明

NBA 常规赛数据来源于 stat-nba(http://www.stat-nba.com),通过数据查询器抓取了这两个赛季中 30 支 NBA 球队的场均数据,数据的信息如表 5-3 所示。

表 5-3 数据的信息

序　号	SEQ
球队名称	TEAM
赛季	SEASON
投篮命中率	FG
投篮命中数	FGM
投篮出手次数	FGA
三分球命中率	3P
三分球命中数	3PM
三分球出手次数	3PA
罚球命中率	FT
罚球命中次数	FTM
罚球出手次数	FTA
篮板	REBS
前场篮板	OREB
后场篮板	DREB
助攻	AST
抢断	STL
盖帽	BLK
失误	TO
犯规	FOVLS
场均得分	PTS
场均失分	PTLS
胜场	W
负场	L

球队的分区信息如表 5-4 所示。

表 5-4 球队的分区信息

球队名称	TEAM
所属分赛区	DIVISION
所属联盟	UNION

2. 在 Hive 中加载 NBA 常规赛数据，并使用 Spark SQL 查询

1）将 NBA 球队的场均数据信息、NBA 球队分区信息数据文件上传到 HDFS。

2）在 Hive 中创建外部表：NBA 球队的场均数据表（PLAYOFFS_1416）、NBA 球队的分区信息表（TEAMINFO）。

3）使用 Spark SQL 统计 NBA 常规赛各类查询数据。

（1）将数据文件上传到 HDFS。

篮球参考网站（www.basketball-reference.com）提供了 NBA 篮球历届比赛球员、球队、季节赛、领先者、分数、季后赛、篮球指数的详细数据。我们可以从网站上（http://www.basketball-reference.com/leagues/NBA_2017_totals.html）下载 NBA 篮球运动员历史数据。将数据文件上传到 HDFS 指定的目录下：

```
hadoop dfs -put ./playOffs1416.csv /library/SparkSQL/function/nba/playoffs1416
hadoop dfs -put ./teamInfo.csv /library/SparkSQL/function/nba/teaminfo
```

（2）启动 Spark-Shell，根据上述信息分别创建外部表。

```
scala > hiveContext.sql("create external table PLAYOFFS_1416(SEQ INT,TEAM STRING,SEASON STRING,FG DOUBLE,FGM DOUBLE,FGA DOUBLE,3P DOUBLE,3PM DOUBLE,3PA DOUBLE,FT DOUBLE,FTM DOUBLE,FTA DOUBLE,REBS DOUBLE,OREB DOUBLE,DREB DOUBLE,AST DOUBLE,STL DOUBLE,BLK DOUBLE,TO DOUBLE,FOVLS DOUBLE,PTS DOUBLE,PTLS DOUBLE,W INT,L INT) ROW FORMAT DELIMITED FIELDS TERMINATED BY ','LINES TERMINATED BY '\n'STORED AS TEXTFILE LOCATION
    '/library/SparkSQL/function/nba/playoffs1416'")
......

scala > hiveContext.sql(" create external table TEAMINFO(TEAM STRING, DIVISION STRING, UNION
    STRING) ROW FORMAT DELIMITED FIELDS TERMINATED BY ','LINES TERMINATED BY
    '\n'STORED AS TEXTFILE LOCATION '/library/SparkSQL/function/nba/teaminfo'")
......
```

在 Spark-Shell 中，查询 Hive 数据库 NBA 球队的场均数据表（PLAYOFFS_1416）、NBA 球队的分区信息表（TEAMINFO）中的数据记录数：

```
scala > hiveContext.sql("select count(*) from PLAYOFFS_1416").show
+-----+
|  _c0|
+-----+
|   60|
+-----+

scala > hiveContext.sql("select count(*) from TEAMINFO").show
```

```
+-----+
| _c0 |
+-----+
|  30 |
+-----+
```

（3）使用 Spark SQL 统计 NBA 常规赛各类查询数据。

1）场均数据和球队分区信息关联查询。

对过去的两个赛季，每支球队都具有自己的场均指标，我们可以根据胜负场计算出各支球队的胜率，为了更好地、清晰地进行指标分析，我们将场均数据和球队的分区信息进行关联，保存为一个 dataFrame，代码如下：

```
scala > val dataFrame = hiveContext.sql("SELECT tmp.*,round(W/(W+L),4)*100 AS
    WR,info.DIVISION,info.UNION FROM PLAYOFFS_1416 tmp JOIN TEAMINFO info ON
    tmp.TEAM = info.TEAM");
```

给这个 DataFrame 注册临时表，后续的统计我们要运用这张临时表：

```
scala > dataFrame.registerTempTable("PG_DATA");
```

查询 dataFrame 的记录数：

```
scala > dataFrame.count( )
…
16/04/24 18:58:10 INFO scheduler.DAGScheduler:Job 10 finished:count at <console>:32,took
    4.206738 s
res11:Long = 60
```

dataFrame 的记录数是 60，即将 NBA 球队的场均数据表（PLAYOFFS_1416）、NBA 球队的分区信息表（TEAMINFO）根据球队名称关联以后，共查询的记录数为 60。dataFrame 每一行的记录包括球队名称、赛季、投篮命中率、投篮命中数、投篮出手次数、三分球命中率、三分球命中数、三分球出手次数、罚球命中率、罚球命中次数、罚球出手次数、篮板、前场篮板、后场篮板、助攻、抢断、盖帽、失误、犯规、场均得分、场均失分、胜场、负场、胜率 WR（即胜场/（胜场+负场），结果四舍五入后保留 4 位小数）、所属分赛区、所属联盟等信息。

我们查询近两个赛季金州勇士队的情况，从 DataFrame 注册的临时表 PG_DATA 中查询金州勇士队在 NBA 14~15 赛季、NBA 15~16 赛季比赛记录的相关内容，包括：球队名称、赛季、胜率。

```
hiveContext.sql("SELECT TEAM,SEASON,WR FROM PG_DATA WHERE TEAM = 'Golden State
    Warriors '").show
…
```

第5章 Spark SQL内置函数与窗口函数

```
+------------------+----------+-------------------+
|              TEAM|    SEASON|                 WR|
+------------------+----------+-------------------+
|  Golden StateWarr...|   14-15|  81.71000000000001|
|  Golden StateWarr...|   15-16|              89.02|
+------------------+----------+-------------------+
```

2)根据胜负场次统计两个赛季能够进入季后赛的球队:按照赛季、所属联盟分组,对胜场场次、投篮命中率进行降序排列,使用窗口函数 row_number() 查询出球队的排名情况。

```
scala > val playoffsRankDF = hiveContext.sql("SELECT TEAM,SEASON,UNION,W,L,WR,FG,
    row_number() OVER(PARTITION BY SEASON,UNION ORDER BY W DESC,FG DESC) as rn
    FROM PG_DATA");
```

playoffsRankDF 查询结果每行记录的内容包括:球队名称、赛季、所属联盟、胜场、负场、胜率、投篮命中率、使用 row_number() 函数查询球队排名 rn 等信息。

将 playoffsRankDF 注册为临时表 PLAYOFFSRANK:

```
scala > playoffsRankDF.register TempTable("PLAYOFFSRANK");
```

查询 NBA 15-16 赛季东西部联盟进入季后赛的球队信息的具体步骤如下:
① 查询 playoffsRankDF 注册的临时表 PLAYOFFSRANK;
② 根据球队前8强排名 rn、赛季 SEASON、所属联盟 UNION 等条件进行查询。
③ 查询 NBA 15~16 赛季西部联盟前8强的球队信息,包括球队名称 TEAM、胜场 W、负场 L、球队排名 rn。
④ 查询 NBA 15~16 赛季东部联盟前8强的球队信息,包括球队名称 TEAM、胜场 W、负场 L、球队排名 rn。

在上面的结果中分别查找 2015~2016 赛季东西部进入季后赛的球队信息。
NBA15~16 赛季西部前8强查询结果如下:

```
scala > hiveContext.sql("SELECT TEAM,W,L,rn FROM PLAYOFFSRANK WHERE rn < 9 AND
    SEASON = '15-16' AND UNION = 'WEST'").show
......
+------------------+----+---+---+
|              TEAM|   W|  L| rn|
+------------------+----+---+---+
|  Golden StateWarr...|  73|  9|  1|
|   San Antonio Spurs|  67| 15|  2|
|  Oklahoma CityThu...|  55| 27|  3|
|  Los Angeles Clippers|  53| 29|  4|
|   Portland Trail Bl...|  44| 38|  5|
```

```
|     Dallas Mavericks | 42 | 40 | 6 |
|    Memphis Grizzlies | 42 | 40 | 7 |
|     Houston Rockets  | 41 | 41 | 8 |
+----------------------+----+----+---+
```

NBA15~16赛季东部联盟前8强查询结果如下：

```
scala > hiveContext.sql("SELECT TEAM,W,L,rn FROM PLAYOFFSRANK WHERE rn < 9 AND
         SEASON = '15 – 16 'AND UNION = 'EAST '").show
……
+-------------------+----+----+----+
|              TEAM |  W |  L | rn |
+-------------------+----+----+----+
| Cleveland Cavaliers| 57 | 25 | 1 |
|    Toronto Raptors | 56 | 26 | 2 |
|        Miami Heat  | 48 | 34 | 3 |
|      Atlanta Hawks | 48 | 34 | 4 |
|     Boston Celtics | 48 | 34 | 5 |
|  Charlotte Hornets | 48 | 34 | 6 |
|    Indiana Pacers  | 45 | 37 | 7 |
|    Detroit Pistons | 44 | 38 | 8 |
+-------------------+----+----+----+
```

3) 针对最近两个赛季常规赛排名，我们可以根据排名的变化，找出东、西部进步幅度最大的球队。首先查找出每支球队的排名涨幅。

具体步骤如下：

① 子查询 T1：NBA 15~16 赛季中 PLAYOFFSRANK 临时表中的记录，内容包括：球队名称、赛季、所属联盟、胜场、负场、胜率、投篮命中率、使用 row_number( ) 函数查询球队排名 rn。

② 子查询 T2：NBA 14~15 赛季中 PLAYOFFSRANK 临时表中的记录，内容包括：球队名称、赛季、所属联盟、胜场、负场、胜率、投篮命中率、使用 row_number( ) 函数查询球队排名 rn。

③ 将子查询 T1 和子查询 T2 根据球队名称进行关联。

④ 从关联表中查询 NBA 球队排名涨幅差异记录结果，内容包括：NBA 14~15 赛季球队名称、NBA 14~15 赛季球队所属联盟、NBA 14~15 赛季排名 LAST_RANK、NBA 15~16 赛季排名 CURRENT_RANK、NBA 15~16 赛季排名和 NBA 14~15 赛季排名相减即排名涨幅差异 DIFF 等信息。

⑤ 将查询结果注册临时表 RANKDIFFDF。

⑥ 使用 rank 函数对排名涨幅差异 DIFF 字段再次进行涨幅幅度排名，分别查询出东部联盟、西部联盟排名涨幅第一的球队。

第5章 Spark SQL内置函数与窗口函数

NBA球队排名涨幅差异记录查询：

```
scala > val rankDiffDF = hiveContext.sql("SELECT T2.TEAM,T2.UNION,T2.rn AS LAST_RANK,
    T1.rn AS CURRENT_RANK,(T2.rn – T1.rn) AS DIFF FROM(SELECT * FROM
    PLAYOFFSRANK WHERE SEASON = '15 – 16 ')T1 JOIN(SELECT * FROM PLAYOFFSRANK
    WHERE SEASON = '14 – 15 ')T2 ON T1.TEAM = T2.TEAM");
```

将rankDiffDF注册临时表RANKDIFFDF：

```
scala > rankDiffDF.registerTempTable("RANKDIFFDF");
```

使用rank函数对排名涨幅差异DIFF字段再次进行涨幅幅度排名：

① 从临时表RANKDIFFDF进行子查询，根据球队所属联盟分组，按照排名涨幅差异DIFF字段进行降序排列，使用rank()窗口函数对排名涨幅情况进行排名，查询记录内容包括：NBA 14~15赛季球队名称、NBA 14~15赛季球队所属联盟、NBA 14~15赛季排名LAST_RANK、NBA 15~16赛季排名CURRENT_RANK、NBA 15~16赛季排名 – NBA 14~15赛季排名（即排名涨幅差异DIFF）排名涨幅的排名rn。

② 查询出东部联盟、西部联盟排名涨幅第一（t.rn = 1）的球队，内容包括：球队所属联盟、球队名称。

```
scala > hiveContext.sql("SELECT t.UNION,t.TEAM FROM(SELECT tmp.*,rank()OVER
    (PARTITION BY UNION ORDER BY DIFF DESC)AS rn FROM RANKDIFFDF tmp)t
    WHERE t.rn = 1").show
……
+------+----------------+
| UNION|            TEAM|
+------+----------------+
|  WEST| Oklahoma CityThu...|
|  EAST|      Miami Heat|
+------+----------------+
```

从统计的结果可以看出，较上个赛季而言，本赛季东、西部进步最快的球队是西部的雷霆队和东部的热火队。

4）分析统计出本赛季命中率、三分球命中率、防守能力最强的三支球队。

具体步骤如下：

① 根据投篮命中率FG进行降序排序，使用窗口函数rank()对投篮命中率进行排名，作为FG_RANK字段。

② 根据三分球命中率3P进行降序排序，使用窗口函数rank()对三分球命中率进行排名，作为3P_RANK字段。

③ 根据场均失分PTLS进行升序排序，使用窗口函数rank()对场均失分最少进行排名，作为PTLS_RANK字段。

④ 从 NBA 球队的场均数据表（PLAYOFFS_1416）中查询 NBA 15~16 赛季的记录，内容包括：球队名称、投篮命中率排名、三分球命中率排名、场均失分排名。
⑤ 查询投篮命中率最高的球队，即投篮命中率排名第一的球队（FG_RANK=1）。
⑥ 查询三分球命中率最高的球队，即三分球命中率排名第一的球队（3P_RANK=1）。
⑦ 查询防守能力最强的球队，即场均失分最少排名第一的球队（PTLS_RANK=1）。

```
scala > val dataRankDF = hiveContext. sql( "SELECT TEAM,rank( )OVER(ORDER BY FG DESC)AS
    FG_RANK,rank( )OVER(ORDER BY 3P DESC)AS 3P_RANK,rank( )OVER(ORDER BY
    PTLS ASC)AS PTLS_RANK FROM PLAYOFFS_1416 WHERE SEASON = '15 – 16 '")
16/04/24 23:49:36 INFO parse. ParseDriver:Parsing command:SELECT TEAM,rank( )OVER
    (ORDER BY FG DESC)AS FG_RANK,rank( )OVER(ORDER BY 3P DESC)AS 3P_RANK,
    rank( )OVER(ORDER BY PTLS ASC)AS PTLS_RANK FROM PLAYOFFS_1416 WHERE
    SEASON = '15 – 16 '
……
```

将 dataRankDF 注册临时表 DATARANK。

```
scala > dataRankDF. registerTempTable( "DATARANK" ) ;
```

查询投篮命中率最高的球队：

```
scala > hiveContext. sql( "SELECT * FROM DATARANK WHERE FG_RANK = 1" ). show
……
+---------------+---------+---------+---------+
|           TEAM| FG_RANK | 3P_RANK |PTLS_RANK|
+---------------+---------+---------+---------+
|Golden StateWarr...|     1 |       1 |      19 |
+---------------+---------+---------+---------+
```

查询三分球命中率最高的球队：

```
scala > hiveContext. sql( "SELECT * FROM DATARANK WHERE 3P_RANK = 1" ). show
……
+---------------+---------+---------+---------+
|           TEAM| FG_RANK | 3P_RANK |PTLS_RANK|
+---------------+---------+---------+---------+
|Golden StateWarr...|     1 |       1 |      19 |
+---------------+---------+---------+---------+
```

查询防守能力最强的球队：

```
scala > hiveContext. sql( "SELECT * FROM DATARANK WHERE PTLS_RANK = 1" ). show
……
```

```
+----------------------+---------+---------+-----------+
|        nn    TEAM    | FG_RANK | 3P_RANK | PTLS_RANK |
+----------------------+---------+---------+-----------+
|   San Antonio Spurs  |    2    |    2    |     1     |
+----------------------+---------+---------+-----------+
```

从这 3 个指标来看本赛季联盟投篮命中率、三分球命中率排名第一的均为勇士队，但防守能力处于中下水平，而防守能力排名第一的马刺队，在投篮命中率方面均为联盟第二，可见马刺队在攻防能力上均有较高的水准。

5.3 本章小结

通过本章的学习，读者可以掌握 Spark SQL 中内置函数和窗口函数使用的基本流程，并学会通过这些函数对业务数据进行分析；同时也可以掌握以 Hive 作为数据仓库、Spark SQL 作为计算引擎的数据统计分析方法。通过本章的案例，窗口函数的使用非常重要，可以通过它解决诸如 Top N 排序等较为复杂的查询，给开发者提供了极大的便利。由于本章篇幅有限，其他的内置函数可以参照 Spark 官方文档进行学习使用。

第 6 章 Spark SQL UDF 与 UDAF

6.1 UDF 概述

用户自定义函数（User Define Function，UDF），用户可以将自己编写的自定义函数（UDF）加入到用户会话中（交互式查询或者通过脚本执行），UDF 将和内置的函数一样使用。

Spark SQL 中 org.apache.spark.sql.functions.scala 提供了大量的内置函数，包括聚合函数、日期时间函数、排序函数、非聚合函数、数学函数、窗口函数、字符串函数、集合函数等，例如：使用 count、max、min、avg 等内置函数计算计数、最大值、最小值、平均值等各种查询统计。但 Spark SQL 支持的 SQL 仍然有限，有时候 Spark SQL 提供的函数功能不能满足业务需要，就需要用户进行自定义函数（UDF）开发。Spark SQL 的 UDF 类型包括普通 UDF、用户自定义聚集函数 UDAF 等。其中 UDF 操作单行数据记录，产生单行数据输出；UDAF 接受多条数据记录输入，产生单行数据输出。

6.2 UDF 示例

编写 UDF 分为三步：

第一步：编写 UDF 函数。根据业务需求，编写自定义化的 UDF 函数。

第二步：注册 UDF 函数。通过 SQLContext 注册 UDF，在 Scala 2.10.x 版本中，UDF 函数最多可以接受 22 个输入参数。

第三步：写 SQL 语句。直接在 SQL 语句中使用 UDF，就像使用 SQL 自动的内部函数一样。

本章用户自定义函数（UDF）示例包括：Hobby_count 函数，将字符串以","切分，计算切分后数组的长度；Combine 函数，合并 s1，s2 两个字符串，如果 s1 为空，返回 s2，否则返回 s1＋s2；str2Int 函数，将字符串转为整数；Wsternstate 函数，判断给定的字符串是否包含在" CA"" OR" " WA" " AK"中；manyCustomers 函数，判断给定的数是否大于 2；stateRegion 函数，模式匹配 state 进行处理；discountRatio 函数，求折扣和原价的比值；makeStruct 函数，将 sales，discounts 组合成 SalesDiscount 结构；myDateFilter 函数，过滤出 8 月份的记录；makeDT 函数，将 3 个字符串合并连接。各示例的数据来源来自程序生成的模拟数据。

第6章 Spark SQL UDF与UDAF

6.2.1 Hobby_count 函数

函数名称：hobby_count。

函数功能：将字符串以","切分，然后求切分后形成的数组的长度。

函数示例：

1）定义 NameHobbies 的 case class 类，构建 NameHobbies 类型的 Seq 集合变量 data，通过 sc 的 parallelize 方法读入 data 数据，调用 toDF() 方法转换成 DataFrame，然后注册成临时表 NameHobbiesTable。

2）在 sqlContext.udf 中注册 hobby_count 的自定义函数，将字符串以","切分，求切分后形成的数组的长度。

3）在临时表 NameHobbiesTable 中执行查询操作，打印出结果。

```scala
import org.apache.spark.{SparkConf,SparkContext}
import org.apache.spark.sql.SQLContext

//继承 App,免写 main 方法
object MyUDF extends App{
  //设置 Spark 运行时的配置参数,setAppName("MyUDF")表示将应用程序的名字设为 MyUDF,
  setMaster("local[*]")   //Spark 在本地多线程运行(指定所有可用内核)
  val conf = new SparkConf().setAppName("MyUDF").setMaster("local[*]")

  //建立 Spark 上下文,这是通向 Spark 集群的唯一入口
  val sc = new SparkContext(conf)

  //建立基于 Spark 上下文的 SQL 上下文
  val sqlContext = new SQLContext(sc)

  //如果不导入,后面 toDF()时会报错
  import sqlContext.implicits._

  case class NameHobbies(name:String,hobbies:String)
  val data = Seq(NameHobbies("sasuke","jog,code,cook"),
  NameHobbies("naruto","travel,dance"))

  //加载数据,将其注册为表,表名为:NameHobbiesTable
  sc.parallelize(data).toDF().registerTempTable("NameHobbiesTable")

  //注册名为 hobby_count 的自定义函数,这个自定义函数的功能是:将字符串以",",切分,然后求切分后形成的数组的长度
  sqlContext.udf.register("hobby_count",(s:String) => s.split(',').size)
```

```
        //执行查询操作,将 hobby_count(hobbies)的结果列重命名为 hobby_count,在 Driver 上打印
    结果
    sqlContext.sql("select * ,hobby_count(hobbies) as hobby_count from NameHobbiesTable").show()

        //执行清理操作,清除中间数据,停止 Spark 上下文
        sc.stop()
    }
```

在本地运行,结果如下所示。

```
+------+------------+-----------+
| name |   hobbies  |hobby_count|
+------+------------+-----------+
|sasuke|jog,code,cook|     3    |
|Naruto| travel,dance|     2    |
+------+------------+-----------+
```

6.2.2　Combine 函数

函数名称:combine。

函数功能:如果 s1 为空,返回 s2,否则返回 s1 + s2。

函数示例:

1) 定义 NullTable 的 case class 类,构建 NullTable 类型的 Seq 集合变量 data,通过 sc 的 parallelize 方法读入 data 数据,调用 toDF() 方法转换成 DataFrame,注册成临时表 NullTable。

2) 在 sqlContext.udf 中注册 combine 的自定义函数,如果 s1 为空,返回 s2,否则返回 s1 + s2。

3) 在临时表 NullTable 中执行查询操作,打印出结果。

```
import org.apache.spark.{SparkConf,SparkContext}
import org.apache.spark.sql.SQLContext
object MyUDF extends App{
    val conf = new SparkConf().setAppName("MyUDF").setMaster("local[*]")
    val sc = new SparkContext(conf)
    val sqlContext = new SQLContext(sc)

    import sqlContext.implicits._

    case class NullTable(str1:String,str2:String)
    val data = Seq(NullTable(null,"123"),NullTable("123","456"))
```

第6章 Spark SQL UDF与UDAF

```
        //加载数据,将其注册为表,表名为:NullTable
        sc.parallelize(data).toDF().registerTempTable("NullTable")

        //注册名为 combine 的自定义函数,这个自定义函数的功能是:如果 s1 为空,返回 s2,否则返回
s1 + s2
        sqlContext.udf.register("combine", (s1:String,s2:String) => {if(s1 == null)s2 else s1 + s2})

        //执行查询操作,将 combine(str1,str2)的结果列重命名为 AB,在 Driver 上打印结果
        sqlContext.sql("select combine(str1,str2) as AB from NullTable").show()

        sc.stop()
}
```

在本地运行,结果如下所示。

```
+------+
|    AB|
+------+
|   123|
|123456|
+------+
```

6.2.3 Str2Int 函数

函数名称:str2Int。

函数功能:将字符串转为整数。

函数示例:

1)定义 Simple 的 case class 类,构建 Simple 类型的 Seq 集合变量 data,通过 sc 的 parallelize 方法读入 data 数据,调用 toDF() 方法转换成 DataFrame,注册成临时表 Simple。

2)在 sqlContext.udf 中注册 str2Int 的自定义函数,将字符串转为整数。

3)在临时表 Simple 中执行查询操作,将 str2Int(str)的结果列重命名为 str2Int,查询打印结果。

4)在临时表 Simple 中执行查询操作,cast 为 SparkSQL 的内置函数,查询打印结果。

```
import org.apache.spark.{SparkConf,SparkContext}
import org.apache.spark.sql.SQLContext

objectMyUDF extends App{
    val conf = newSparkConf().setAppName("MyUDF").setMaster("local[*]")
    val sc = newSparkContext(conf)
    val sqlContext = new SQLContext(sc)
```

```
        import sqlContext.implicits._
        case class Simple(str:String)
        val data = Seq(Simple("12"),Simple("34"),Simple("56"),Simple("78"))

        //加载数据,将其注册为表,表名为:Simple
        sc.parallelize(data).toDF().registerTempTable("Simple")

        //注册名为 str2Int 的自定义函数,这个自定义函数的功能是:将字符串转为整数
        sqlContext.udf.register("str2Int",(s:String) => s.toInt)

        //执行查询操作,将 str2Int(str)的结果列重命名为 str2Int,在 Driver 上打印结果
        sqlContext.sql("SELECT str2Int(str) AS str2Int FROM Simple").show()

        //执行查询操作,在 Driver 上打印结果,其中 cast 为 SparkSQL 的内置函数
        sqlContext.sql("SELECT cast(str AS Int) FROM Simple").show()

        sc.stop()
    }
```

在本地运行,结果如下所示。

```
+-----+
|str2Int|
+-----+
|   12|
|   34|
|   56|
|   78|
|   90|
+-----+


+---+
|str|
+---+
| 12|
| 34|
| 56|
| 78|
| 90|
+---+
```

第6章 Spark SQL UDF与UDAF

6.2.4 Wsternstate 函数

函数名称：westernState，查询顾客是否归属于美国加利福尼亚州 CA（California）、俄勒冈州 OR（Oregon）、华盛顿州 WA（Washington）、阿拉斯加州 AK（Alaska）。

函数功能：查询统计顾客商品的购买记录，顾客信息包括 ID，姓名，销售额，折扣金额，所在州，其中 CA 是美国加利福尼亚州（California）、AZ 是亚利桑那州（Arizona）、MA 是马萨诸塞州（Massachusetts）。本示例要统计查询四个州的顾客购买记录（美国加利福尼亚州 CA（California）、俄勒冈州 OR（Oregon）、华盛顿州 WA（Washington）、阿拉斯加州 AK（Alaska））。通过自定义函数 westernState，传入顾客所在州的参数，由函数 westernState 判断是否包含在这四个州中，即给定的字符串是否包含在"CA""OR" "WA" "AK"中。

函数示例：

1）定义 Customer 的 case class 类，构建 Customer 类型的 Seq 集合变量 custs，通过 sc 的 parallelize 方法读入 custs 数据，调用 toDF()方法转换成 DataFrame，注册成临时表 customerTable。

2）在 sqlContext.udf 中注册 westernState 的自定义函数，westernState_是偏函数，westernState 函数中传入一个参数，判断参数给定的字符串是否包含在"CA""OR""WA""AK"中。

3）在临时表 customerTable 中执行查询操作，自定义函数作为过滤条件（传入的 state 是否在"CA","OR","WA","AK"中），查询打印结果。

```scala
import org.apache.spark.{SparkConf,SparkContext}
import org.apache.spark.sql.SQLContext

object MyUDF extends App{
  val conf = newSparkConf().setAppName("MyUDF").setMaster("local[*]")
  val sc = newSparkContext(conf)
  val sqlContext = new SQLContext(sc)

  import sqlContext.implicits._

  case class Customer(id:Integer,name:String,sales:Double,discounts:Double,state:String)
  val custs = Seq(Customer(1,"Widget Co",120200.00,0.00,"AZ"),
                  Customer(2,"Acme Widgets",410600.00,560.00,"CA"),
                  Customer(3,"Widgetry",410550.00,230.00,"CA"),
                  Customer(4,"Widgets R Us",410505.00,0.0,"CA"),
                  Customer(5,"YeOldeWidgete",500.00,0.0,"MA"))

  //加载数据,将其注册为表,表名为:customerTable
  sc.parallelize(custs).toDF().registerTempTable("customerTable")

  //这个函数的功能是:确定给定的字符串是否包含在"CA","OR","WA","AK"中
```

```
        def westernState(state:String) = Seq("CA","OR","WA","AK").contains(state)

        //注册名为 westernState 的自定义函数,注意 westernState 与_之间存在一个空格,这在 Scala 的
        语法中被称作偏函数(partially function)
        sqlContext.udf.register("westernState",westernState _)

        //执行查询操作,自定义函数在这里作为过滤条件,在 Driver 上打印结果
        sqlContext.sql("SELECT * FROM customerTable WHERE westernState(state)").show()

        sc.stop()
    }
```

在本地运行,结果如下所示。

查询顾客归属于美国加利福尼亚州 CA(California)、俄勒冈州 OR(Oregon)、华盛顿州 WA(Washington)、阿拉斯加州 AK(Alaska)这四个州的顾客商品购买记录,这里亚利桑那州 AZ(Arizona)、马萨诸塞州 MA(Massachusetts)不在这四个州范围里面,查询出归属于美国加利福尼亚州 CA 的所有顾客商品购买记录。

```
+---+------------+--------+---------+-----+
| id|        name|   sales|discounts|state|
+---+------------+--------+---------+-----+
|  2| Acme Widgets|410600.0|    560.0|   CA|
|  3|     Widgetry|410550.0|    230.0|   CA|
|  4| Widgets R Us|410505.0|      0.0|   CA|
+---+------------+--------+---------+-----+
```

6.2.5 ManyCustomers 函数

函数名称:manyCustomers。

函数功能:判断给定的数是否大于 2。

函数示例:

1)定义 manyCustomers 函数,判断给定的数是否大于 2。

2)在 sqlContext.udf 中注册 manyCustomers 的自定义函数,manyCustomers_是偏函数,manyCustomers 函数中传入一个参数,判断给定的数是否大于 2。

3)在临时表 customerTable 中执行查询操作,根据州名分组,查询所属哪个州、顾客购买次数,HAVING 子句调用自定义函数 manyCustomers 查询顾客购买次数大于 2 次的记录,统计查询打印结果。

```
        def manyCustomers(cnt:Long) = cnt > 2
```

第6章 Spark SQL UDF与UDAF

```
    //注册名为 manyCustomers 的自定义函数,注意 manyCustomers 与_之间存在一个空格,这在
    Scala 的语法中被称作偏函数(Partially Function)
    sqlContext.udf.register("manyCustomers",manyCustomers _)

    //执行查询操作,自定义函数此处作用于分组后的数据,在 Driver 上打印出结果
    sqlContext.sql(
      s"""
        |SELECT state,COUNT(id) AScustCount
        |FROMcustomerTable
        |GROUP BY state
        |HAVINGmanyCustomers(custCount)
    """.stripMargin).show()
```

其中,s"""""".stripMargin 语法,在写复杂的 SQL 语句时很有用,可以写出层次分明的 SQL 语句,在本地运行,结果如下所示。

```
+---+---------+
|state|custCount|
+---+---------+
| CA |        3|
+---+---------+
```

6.2.6 StateRegion 函数

函数名称:stateRegion。
函数功能:使用模式匹配实现不同的 state 的匹配处理。
函数示例:
1)定义 stateRegion 函数,模式匹配给定的参数属于哪个区域:西部、东北、西南。
2)在 sqlContext.udf 中注册 stateRegion 的自定义函数,stateRegion_是偏函数,stateRegion 函数中传入一个参数,判断给定的参数属于哪个地区。
3)在临时表 customerTable 中执行查询操作,根据州所属区域分组,查询销售总金额、调用 stateRegion 函数查询州所属区域,统计查询打印结果。

```
    //这个自定义函数使用模式匹配实现
    def stateRegion(state:String) = state match {
      case "CA" | "AK" | "OR" | "WA" => "West"
      case "ME" | "NH" | "MA" | "RI" | "CT" | "VT" => "NorthEast"
      case "AZ" | "NM" | "CO" | "UT" => "SouthWest"
    }
```

```
         //注册名为 stateRegion 的自定义函数,注意 stateRegion 与_之间存在一个空格,这在 Scala 的语
法中被称作偏函数(Partially Function)
sqlContext.udf.register("stateRegion",stateRegion _)

         //执行查询操作,此处自定义函数被用作分组条件,将 stateRegion(state)的结果列重命名为
Region,在 Driver 上打印出结果
sqlContext.sql(
    s"""
        |SELECT SUM(sales) AS totalSales,stateRegion(state) AS Region
        |FROM customerTable
        |GROUP BY stateRegion(state)
""".stripMargin).show()
```

在本地运行,结果如下所示。

```
+---------+---------+
|totalSales|   Region|
+---------+---------+
|1231655.0|     West|
|    500.0| NorthEast|
| 120200.0| SouthWest|
+---------+---------+
```

6.2.7 DiscountRatio 函数

函数名称:discountRatio。

函数功能:求折扣和原价的比值。

函数示例:

1)定义 discountRatio 函数,两个入参分别为销售额、折扣金额,discountRatio 函数计算折扣率,即折扣金额与销售额的折扣百分比。

2)在 sqlContext.udf 中注册 discountRatio 的自定义函数,discountRatio_是偏函数,discountRatio 函数中传入销售额、折扣金额,计算折扣率。

3)在临时表 customerTable 中执行查询操作,查询 ID 序号、折扣率(折扣金额/销售额),统计查询打印结果。

```
def discountRatio(sales:Double,discounts:Double) = discounts / sales

         //注册名为 discountRatio 的自定义函数,注意 discountRatio 与_之间存在一个空格,这在 Scala
的语法中被称作偏函数(partially function)
sqlContext.udf.register("discountRatio",discountRatio _)
```

第6章 Spark SQL UDF与UDAF

```
    //执行查询操作,将 discountRatio(sales,discounts)的结果列重命名为ratio,在 Driver 上打印结果
    sqlContext. sql(
      s"""
        |SELECT id,discountRatio(sales,discounts) AS ratio
        |FROM customerTable
      """. stripMargin). show( )
```

在本地运行,结果如下所示。

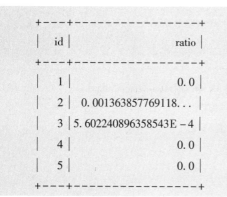

6.2.8 MakeStruct 函数

函数名称:makeStruct。

函数功能:从顾客商品购买记录表 CustomerTable 中查询销售额、折扣金额信息,将销售金额及折扣金额组合在一起显示,将 sales 和 discounts 组合成 SalesDiscount 结构。

函数示例:

1) 定义 SalesDiscount case class,其成员变量为销售金额、折扣金额。

2) 定义 makeStruct 函数,传入销售额、折扣金额,将销售额、折扣金额组合成一个 case class,将 SalesDiscount case class 作为函数结果返回。

3) 在 sqlContext. udf 中注册 makeStruct 的自定义函数,makeStruct_是偏函数,makeStruct 函数中传入销售额、折扣金额,返回 case class 类。

3) 在临时表 customerTable 中执行查询操作,查询(销售金额、折扣金额)结构体,统计查询打印结果。

4) 在临时表 customerTable 中执行查询,查询 ID 序号,(销售金额、折扣金额)结构体 sd,子查询结果为 d;然后再从子查询结果 d 中查询 ID 序号,(销售金额、折扣金额)结构体 sd,统计查询打印结果。

```
case class SalesDiscount( sales:Double,discounts:Double)

def makeStruct( sales:Double,discounts:Double) = SalesDiscount( sales,discounts)
```

```
//注册名为 makeStruct 的自定义函数,注意 makeStruct 与_之间存在一个空格,这在 Scala 的语
法中被称作偏函数(Partially Function)
sqlContext.udf.register("makeStruct",makeStruct _)

//执行查询操作,将 makeStruct(sales,discounts)的结果列重命名为 sd,在 Driver 上打印结果
sqlContext.sql("SELECT makeStruct(sales,discounts) AS sd FROM customerTable").show()

//嵌套查询,s"""""".stripMargin 使用"1"作为连接符,创建多行字符串。
sqlContext.sql(
    s"""
      |SELECT id,sd FROM
      |(
      |SELECT id,makeStruct(sales,discounts) AS sd FROM customerTable
      |) AS d
    """.stripMargin).show()
```

在本地运行,结果如下所示。

```
+-------------+
|           sd|
+-------------+
| [120200.0,0.0]|
|[410600.0,560.0]|
|[410550.0,230.0]|
| [410505.0,0.0]|
|    [500.0,0.0]|
+-------------+

+---+-------------+
| id|           sd|
+---+-------------+
|  1| [120200.0,0.0]|
|  2|[410600.0,560.0]|
|  3|[410550.0,230.0]|
|  4| [410505.0,0.0]|
|  5|    [500.0,0.0]|
+---+-------------+
```

6.2.9　MyDateFilter 函数

函数名称:myDateFilter。

第6章 Spark SQL UDF与UDAF

函数功能：查询学生的报名表，如果学生是8月份报名的，则过滤出8月份的记录，并查询统计8月份报名的学生姓名、报名时间。

函数示例：

1）定义 Entry 的 case class 类，其成员变量为名称、日期，构建 Entry 类型的 Seq 集合变量 data，通过 sc 的 parallelize 方法读入 data 数据，调用 toDF() 方法转换成 DataFrame，注册成临时表 entryTable。

2）定义 myDateFilter 函数，传入日期字符串，如日期字符串中包含"-08-"，则返回 true 值。

3）在 sqlContext.udf 中注册 myDateFilter 的自定义函数，myDateFilter_是偏函数，myDateFilter 函数中传入日期，根据日期字符串过滤出8月份的记录。

4）在临时表 entryTable 中执行查询操作，查询日期在8月份的记录，打印出结果。

```
case class Entry(name:String,when:String)

    val date = Seq(Entry("one","2014-01-01"),
                   Entry("two","2014-03-01"),
                   Entry("three","2014-08-01"),
                   Entry("four","2014-08-15"),
                   Entry("five","2014-12-15"))

    //加载数据,将其注册为表,表名为:entryTable
    sc.parallelize(date).toDF().registerTempTable("entryTable")

    def myDateFilter(date:String) = date.contains("-08-")

//注册名为 myDateFilter 的自定义函数,注意 myDateFilter 与_之间存在一个空格,这在 Scala 的语法中被称作偏函数(Partially Function)
sqlContext.udf.register("myDateFilter",myDateFilter _)

    //执行查询操作,自定义函数 myDateFilter 被用作过滤条件,在 Driver 上打印结果
sqlContext.sql("SELECT * FROM entryTable WHERE myDateFilter(when)").show()
```

在本地运行，结果如下所示。

```
+-----+----------+
| name|      when|
+-----+----------+
|three|2014-08-01|
| four|2014-08-15|
+-----+----------+
```

6.2.10 MakeDT 函数

函数名称：makeDT。

函数功能：将3个字符串连接起来，以空格隔开。

函数示例：

1）定义 Purchase 的 case class 类，其成员变量为顾客 ID、购买 ID、日期、时间、时区、金额，构建 Purchase 类型的 Seq 集合变量 pur。

2）导入 spark 的 sqlContext 隐式转换类 import sqlContext.implicits._，用于将一个 RDD 隐式转换为一个 DataFrame。

3）通过 sc 的 parallelize 方法读入 pur 数据，调用 toDF() 方法转换成 DataFrame，注册成临时表 purchaseTable。

4）定义 makeDT 函数，传入三个参数：日期、时间、时区，将日期、时间、时区三个字符串连接起来，以空格分隔。

5）在 sqlContext.udf 中注册 makeDt 的自定义函数，makeDt_是偏函数，makeDt 函数中传入日期、时间、时区，将三者以空格连接成一个字符串。

6）在临时表 purchaseTable 中执行查询操作，查询金额、日期时间时区，打印出结果。

创建 case class Purchase。

```
scala > case class Purchase(customer_id:Int,purchase_id:Int,date:String,time:
    String,tz:String,amount:Double)
defined class Purchase
```

输入数据。

```
scala > val pur = Seq(Purchase(123,234,"2007-12-12","20:50","UTC",500.99),
    |                 Purchase(123,247,"2007-12-12","15:30","PST",300.22),
    |                 Purchase(189,254,"2007-12-13","00:50","EST",122.19),
    |                 Purchase(187,299,"2007-12-12","07:30","UTC",524.37))
pur:Seq[Purchase] = List(Purchase(123,234,2007-12-12,20:50,UTC,500.99),
    Purchase(123,247,2007-12-12,15:30,PST,300.22),
    Purchase(189,254,2007-12-13,00:50,EST,122.19),
    Purchase(187,299,2007-12-12,07:30,UTC,524.37))
```

导入 sqlContext.implicits._ 隐式函数。如果不导入 implicits._，调用方法 toDF() 时会报错。

```
scala > import sqlContext.implicits._
import sqlContext.implicits._
```

加载数据，将其注册为表，表名为：purchaseTable。

```
scala > sc.parallelize(pur).toDF().registerTempTable("purchaseTable")
```

第6章 Spark SQL UDF 与 UDAF

makeDT 函数的功能是：将3个字符串连接起来，以空格隔开。

```
scala > def makeDT(date:String,time:String,tz:String) = s" $ date $ time $ tz"
makeDT:(date:String,time:String,tz:String)String
```

注册名为 makeDT 的自定义函数，注意 makeDT 与_之间存在一个空格，这在 Scala 的语法中被称作偏函数（Partially Function）。

```
scala > sqlContext.udf.register("makeDt",makeDT _)
res1:org.apache.spark.sql.UserDefinedFunction =
    UserDefinedFunction( < function3 > ,StringType,List())
```

执行查询操作，将 makeDT（date，time，tz）的结果列重命名为 datetime，在 Driver 上打印结果。

```
scala > sqlContext.sql("SELECT amount,makeDT(date,time,tz) AS datetime FROM
    purchaseTable").show()
...
16/06/05 20:47:39 INFO scheduler.DAGScheduler:Job 2 finished:show at < console > :38,
    took 0.704467 s
+------+-------------------+
|amount|           datetime|
+------+-------------------+
|500.99|2007 - 12 - 12 20:50 UTC|
|300.22|2007 - 12 - 12 15:30 PST|
|122.19|2007 - 12 - 13 00:50 EST|
|524.37|2007 - 12 - 12 07:30 UTC|
+------+-------------------+
scala > val fmt = "yyyy - MM - dd hh:mm z"
fmt:String = yyyy - MM - dd hh:mm z
```

执行查询操作，其中 unix_timestamp 是 Spark SQL 的内置函数，在 Driver 上打印结果。

```
scala > sqlContext.sql(s"SELECT customer_id,unix_timestamp(makeDt(date,time,tz),
    ' $ fmt ') AS UTime,amount FROM purchaseTable").show()
......
+-----------+----------+------+
|customer_id|     UTime|amount|
+-----------+----------+------+
|        123|1197492600|500.99|
|        123|1197502200|300.22|
|        189|1197525000|122.19|
|        187|1197444600|524.37|
+-----------+----------+------+
```

6.3 UDAF 概述

用户自定义的聚合函数（User Defined Aggregation Function，UDAF），本身作用于数据集合，能够在聚合操作的基础上进行自定义操作。实际上，UDF 会被 Spark SQL 中的 Catalyst 封装成为 Expression，最终会通过 eval 方法来计算输入的数据 Row（注：此处的 Row 和 DataFrame 中的 Row 没有任何关系）。Spark 对 UDF 的支持较早，Spark 1.1 就推出来了，而 UDAF 是 Spark 1.5 左右才出的。晚了好几个版本，说明 UDAF 有其复杂性，因为它有大量的 Aggregation 之类的操作，是对批量的数据集合进行操作。而 UDF 是对一条数据进行操作，即你会有具体的一条输入的数据，具体如何进行操作，就是一个普通的 Scala 函数。实际应用中也是如此，使用 UDAF 时往往是对数据进行分组的，然后操作。理论上讲，通过 UDF 和 UDAF 可以实现任何功能。

一个 UDAF 维护一个聚合缓冲区来存储每组输入数据的中间结果。它为每个输入行更新此缓冲区，一旦处理完所有输入行，基于该聚合缓冲区的值返回结果。一个 UDAF 继承了基类 UserDefinedAggregateFunction 并实现以下 8 个方法：

- inputSchema：inputSchema 返回 StructType。这个 StructType 的各个字段代表了这个 UDAF 的输入参数。
- BufferSchema：BufferSchema 返回 StructType。这个 StructType 的各个字段代表了这个 UDAF 的中间结果的一个值。
- dataType：dataType 表示此 UDAF 的返回值的数据类型。
- deterministic：deterministic 返回一个布尔值，用于表明在给定输入值的前提下，此 UDAF 是否总是生成一组相同的结果。
- initialize：initialize 用于初始化聚集缓冲区（例如 MutableAggregationBuffer）的值。
- update：update 更新用于输入行的聚集缓冲区（例如 MutableAggregationBuffer）。
- merge：merge 用于合并两个聚集缓冲区，并将结果存储到 MutableAggregationBuffer。
- evaluate：evaluate 用于生成这个 UDAF 的最终值。这个值基于每一行的聚合缓冲区的值。

使用 UDAF 有两种方式：第一种，一个 UDAF 的实例可以立即当作函数使用；第二种，用户可以向 Spark SQL 的功能注册表注册 UDAF，然后通过分配的名称调用此 UDAF。

6.4 UDAF 示例

6.4.1 ScalaAggregateFunction 函数

函数名称：ScalaAggregateFunction。

函数功能：统计单笔销售金额超过 500 的记录销售金额累计相加求和。

函数示例：

第6章　Spark SQL UDF与UDAF

1）定义 ScalaAggregateFunction 类继承至 UserDefinedAggregateFunction：
- 重载实现方法 inputSchema：返回 StructType 字段（销售金额，浮点类型），作为 ScalaAggregateFunction 函数的输入参数；
- 重载实现方法 BufferSchema：返回 StructType 字段（大于 500 的销售额求和值，浮点类型），作为 ScalaAggregateFunction 函数的中间结果的值；
- 重载实现方法 update：如果读入每行 input 的第 0 个元素的值（即销售金额）不为空，而且 input 的销售金额值大于 500，则更新输入行的聚集缓冲区（buffer），buffer 更新第 0 个元素值（即大于 500 的销售额求和 sum 值），其累加当前大于 500 的销售金额值；
- 重载实现方法 merge：merge 用于合并两个聚集缓冲区，将第一个缓冲区大于 500 的销售金额求和值加上第二个缓冲区大于 500 的销售金额求和值，并将结果存储到 MutableAggregationBuffer；
- 重载实现方法 initialize：初始化大于 500 的销售额求和值为 0，用于初始化聚集缓冲区（MutableAggregationBuffer）的值；
- 重载实现方法 deterministic：设置 true，在给定输入值的前提下，ScalaAggregateFunction 生成一组相同的结果；
- 重载实现方法 evaluate：ScalaAggregateFunction 函数的最终计算结果为 buffer 的第 0 个元素（即大于 500 的销售额求和 sum 值）。这个值基于每一行的聚合缓冲区的值；
- 重载实现方法 dataType：dataType 表示 ScalaAggregateFunction 函数返回值的类型是浮点类型。

2）定义顾客 Customer 的 case class 类，其成员变量分别为 ID、姓名、销售额、折扣销售额、所在州等信息。

3）构建 SparkContext 以及 SQLContext，导入 spark 的 sqlContext 隐式转换类 import sqlContext.implicits._，用于将一个 RDD 隐式转换为一个 DataFrame。

4）构建 Customer 类型的 Seq 集合变量 custs，通过 sc 的 parallelize 方法读入 custs 数据，调用 toDF（）方法转换成 DataFrame，使用 customerDF.printSchema（）打印出 customerDF 的 Schema 结构，将 customerDF 注册成临时表 customerTable。

5）创建 ScalaAggregateFunction 实例 mysum，在 sqlContext.udf 中注册 mysum 的自定义 UDAF 函数，将大于 500 的销售金额汇总累加。

6）在临时表 customerTable 执行查询操作，根据州分组，查询所属州、调用 ScalaAggregateFunction 函数实例 mysum，传入销售金额值计算大于 500 的销售金额求和的累加值，然后使用 sqlResult.printSchema（）打印结果的 Schema 结构，使用 sqlResult.show（）打印查询结果。

```
import org. apache. spark. {SparkContext,SparkConf}
import org. apache. spark. sql. {SQLContext,Row}
import org. apache. spark. sql. expressions. {MutableAggregationBuffer,
UserDefinedAggregateFunction}
import org. apache. spark. sql. types. {DoubleType,StructType,DataType}
objectmyUDAF{
```

```scala
class ScalaAggregateFunction extends UserDefinedAggregateFunction{
    //输入类型为 Double 类型
    override def inputSchema:StructType = new StructType().add("sales",DoubleType)

    //中间结果类型为 Double
    override def bufferSchema:StructType = new StructType().add("sumLargeSales",DoubleType)

    override def update(buffer:MutableAggregationBuffer,input:Row):Unit = {
        val sum = buffer.getDouble(0)
        if(!input.isNullAt(0)){
            val sales = input.getDouble(0)
            //超过500才更新相应的缓冲区
            if(sales > 500.0){
                buffer.update(0,sum + sales)
            }
        }
    }

    //合并缓冲区
    override def merge(buffer1:MutableAggregationBuffer,buffer2:Row):Unit = buffer1.update(0,buffer1.getDouble(0) + buffer2.getDouble(0))

    //初始化缓冲区
    override def initialize(buffer:MutableAggregationBuffer):Unit = buffer.update(0,0.0)

    //在给定输入值的前提下,此 UDAF 总是生成一组相同的结果
    override def deterministic:Boolean = true

    //此 UDAF 的最终值为 buffer 中 position 为 0 的值
    override def evaluate(buffer:Row):Any = buffer.getDouble(0)

    //此 UDAF 的返回值的数据类型是 DoubleType
    override def dataType:DataType = DoubleType
}
case class Customer(id:Integer,name:String,sales:Double,discounts:Double,state:String)

def main(args:Array[String]){

    val conf = newSparkConf().setAppName("MyUDAF").setMaster("local[*]")
    val sc = newSparkContext(conf)
    val sqlContext = new SQLContext(sc)
```

第6章 Spark SQL UDF与UDAF

```scala
        import sqlContext.implicits._

        val custs = Seq(Customer(1,"Widget Co",120200.00,0.00,"AZ"),
Customer(2,"Acme Widgets",410600.00,560.00,"CA"),
Customer(3,"Widgetry",410550.00,230.00,"CA"),
Customer(4,"Widgets R Us",410505.00,0.0,"CA"),
Customer(5,"YeOldeWidgete",500.00,0.0,"MA"))

        //加载数据,使其成为 DataFrame
        val customerDF = sc.parallelize(custs).toDF()

        //打印出 DataFrame 的 schema
customerDF.printSchema()

        //注册为表,表名为:customerTable
customerDF.registerTempTable("customerTable")

        //实例化 UDAF
val mysum = new ScalaAggregateFunction()

        //向 Spark SQL 的功能注册表注册名为 mysum 的 UDAF
sqlContext.udf.register("mysum",mysum)

        //mysum 作用于按 state 分组后的数据
        val sqlResult = sqlContext.sql(
            s"""
              | SELECT state,mysum(sales) AS bigsales
              | FROM customerTable
              | GROUP BY state
""".stripMargin)
sqlResult.printSchema()
println()
        //在 Driver 上打印出 SQL 语句的执行结果
sqlResult.show()
    }
}
```

在本地运行,结果如下所示。

```
root
 |-- id:integer(nullable = true)
```

```
 |-- name: string (nullable = true)
 |-- sales: double (nullable = false)
 |-- discounts: double (nullable = false)
 |-- state: string (nullable = true)


root
 |-- state: string (nullable = true)
 |-- bigsales: double (nullable = true)


+-----+---------+
|state| bigsales|
+-----+---------+
|   AZ| 120200.0|
|   CA|1231655.0|
|   MA|      0.0|
+-----+---------+
```

6.4.2　GeometricMean 函数

函数名称：GeometricMean。

函数功能：用于求几何平均值。

函数示例：本例子中同时展示了使用 UDAF 的两种方法。

1) 定义 GeometricMean 类继承至 UserDefinedAggregateFunction：

- 重载实现方法 inputSchema：返回 StructType 字段（数值，浮点类型），作为 GeometricMean 函数的输入参数；
- 重载实现方法 BufferSchema：返回 StructType 字段（（计数，长整型），（乘积，浮点型）），作为 GeometricMean 函数的中间结果的值；
- 重载实现方法 dataType：dataType 表示 GeometricMean 函数返回值的类型是浮点类型；
- 重载实现方法 deterministic：设置 true，在给定输入值的前提下，GeometricMean 生成一组相同的结果；
- 重载实现方法 initialize：初始化 buffer 的第 0 个元素即计数值为 0，初始化 buffer 的第 1 个元素即乘积值为 0.0，用于初始化聚集缓冲区（MutableAggregationBuffer）的值；
- 重载实现方法 update：buffer 更新第 0 个元素值即计数值，当前的数值计数为 1 次，计数值就累加 1；buffer 更新第 1 个元素值即乘积值，将 buffer 第 1 个元素的原乘积值乘以读入的每行元素第 0 个元素即数值，将原乘积乘以新的数值作为更新的乘积值；
- 重载实现方法 merge：merge 用于合并两个聚集缓冲区，将第一个缓冲区的第 0 个元素即计数值加上第二个缓冲区的第 0 个元素即计数值，作为合并以后的缓冲区的第 0 个元素即计数值；将第一个缓冲区的第 1 个元素即乘积值乘以第二个缓冲区的第 1 个元

第6章 Spark SQL UDF与UDAF

素即乘积值，作为合并以后的缓冲区的第1个元素乘积值；
- 重载实现方法 evaluate：GeometricMean 函数的最终计算结果为：将 buffer 的第1个元素（即总乘积值）作为底数，buffer 的第0个元素（即总计数值的倒数）做次幂，两者作幂计算即计算出结果：$\sqrt[n]{x_1 x_2 x_3 \cdots x_n}$；

2）构建 SparkContext 以及 SQLContext，导入 spark 的 org. apache. spark. sql. functions. _，将使用 spark sql 强大的内置函数功能。

3）使用 sqlContext. range 创建一个 DataFrame，其包括1列，列名为 id，列中元素类型为 LongType，列的数值范围从11到50（不包括50），数值步长值为1。创建数值集变量 df。

4）创建 GeometricMean 实例 Val gm = new Geometric Mean。

5）用 UDAF 计算几何平均值方法一：GeometricMean UDAF 的实例作函数使用。df 的 DataFrame 根据 id 列分组，使用 agg 函数调用 GeometricMean UDAF 函数 gm，在 GeometricMean 函数 gm 中传入 id 列数值，计算11至49的几何平均值，然后通过 show 方法展示结果。

6）用 UDAF 计算几何平均值方法二：GeometricMean UDAF 在 Spark SQL 注册使用。在 sqlContext. udf 中注册 gm 的自定义 UDAF 函数，然后 df 的 DataFrame 根据 id 列分组，使用 agg 函数调用表达式，使用已在 Spark SQL 注册的 GeometricMean UDAF 函数 gm，在 GeometricMean 函数 gm 中传入 id 列数值，计算11至49的几何平均值，然后 show 方法展示结果。

```
import org. apache. spark. {SparkContext,SparkConf}
import org. apache. spark. sql. {SQLContext,Row}
import org. apache. spark. sql. expressions. {MutableAggregationBuffer,
UserDefinedAggregateFunction}
import org. apache. spark. sql. types. _

objectmyUDAF{
    classGeometricMean extends UserDefinedAggregateFunction{
        //输入数据类型为 Double
        def inputSchema:org. apache. spark. sql. types. StructType = StructType(StructField(" value ",DoubleType)::Nil)

        //中间结果类型一个是 Long，一个是 Double
        defbufferSchema:StructType = StructType( StructField(" count ",LongType)::StructField(" product ",DoubleType)::Nil)

        //返回结果的类型为 Double
        def dataType:DataType = DoubleType

        //在给定输入值的前提下,此 UDAF 总是生成一组相同的结果
        def deterministic:Boolean = true

        def initialize( buffer:MutableAggregationBuffer):Unit = {
```

```
        buffer(0) = 0L
        buffer(1) = 1.0
    }

    def update(buffer:MutableAggregationBuffer,input:Row):Unit = {
        buffer(0) = buffer.getAs[Long](0) + 1
        buffer(1) = buffer.getAs[Double](1) * input.getAs[Double](0)
    }

    def merge(buffer1:MutableAggregationBuffer,buffer2:Row):Unit = {
        buffer1(0) = buffer1.getAs[Long](0) + buffer2.getAs[Long](0)
        buffer1(1) = buffer1.getAs[Double](1) * buffer2.getAs[Double](1)
    }

    def evaluate(buffer:Row):Any = {
math.pow(buffer.getDouble(1),1.toDouble/buffer.getLong(0))
}
    }

    def main(args:Array[String]){
        val conf = newSparkConf().setAppName("MyUDAF").setMaster("local[*]")
        val sc = newSparkContext(conf)
val sqlContext = new SQLContext(sc)

import org.apache.spark.sql.functions._

        val df = sqlContext.range(11,50)
        val gm = new GeometricMean
//一个 UDAF 的实例可以立即当作函数使用
df.groupBy().agg(gm(col("id")).as("GeometricMean")).show()

        //向 Spark SQL 的功能注册表注册名为 gm 的 UDAF
sqlContext.udf.register("gm",gm)

        //然后,通过分配的名称调用此 UDAF
        df.groupBy().agg(expr("gm(id) as GeometricMean")).show()
    }
}
```

在本地运行,结果如下所示。

```
+------------------+
|     GeometricMean|
```

第6章　Spark SQL UDF与UDAF

```
+------------------+
| 27.64711319471532|
+------------------+


+------------------+
|   GeometricMean  |
+------------------+
| 27.64711319471532|
+------------------+
```

6.4.3　CustomMean 函数

函数名称：CustomMean。

函数功能：

用于计算算术平均数；等同于使用 Spark SQL 本身的内置函数 avg 函数求算术平均值的功能。

函数示例：

1）定义 CustomMean 类继承至 UserDefinedAggregateFunction：

- 重载实现方法 inputSchema：返回 StructType 字段（item，浮点类型），作为 CustomMean 函数的输入参数；
- 重载实现方法 BufferSchema：返回 StructType 字段（（求和值 sum，浮点类型），（计数次数，长整型）），作为 CustomMean 函数的中间结果的值；
- 重载实现方法 dataType：dataType 表示 CustomMean 函数返回值的类型是浮点类型；
- 重载实现方法 deterministic：设置 true，在给定输入值的前提下，CustomMean 生成一组相同的结果；
- 重载实现方法 initialize：初始化 buffer 的第 0 个元素即求和值 sum 为 0，初始化 buffer 的第 1 个元素即计数次数值为 0L，用于初始化聚集缓冲区（MutableAggregationBuffer）的值；
- 重载实现方法 update：buffer 更新第 0 个元素值即求和值 sum，读入 input 每行的第 0 个元素 item 的数值，将原求和值 sum 和 item 数值相加作为新的求和值 sum；buffer 更新第 1 个元素值即计数次数累加值，每读入 input 的一行元素，计数次数为 1，将 buffer 第 1 个元素的原计数次数累加值加上 1 作为新的计数次数累加值；
- 重载实现方法 merge：merge 用于合并两个聚集缓冲区，将第一个缓冲区的第 0 个元素即求和值 sum 加上第二个缓冲区的第 0 个元素即求和值 sum，作为合并以后的缓冲区的第 0 个元素即求和值 sum；将第一个缓冲区的第 1 个元素即计数次数累加值加上第二个缓冲区的第 1 个元素即计数次数累加值，作为合并以后的缓冲区的第 1 个元素计数次数累加值；
- 重载实现方法 evaluate：CustomMean 函数的最终计算结果为：将 buffer 的第 0 个

元素即总求和值 sum 除以 buffer 的第 1 个元素即总计数值，计算出算术平均值。
$A_n = \dfrac{a_1 + a_2 + a_3 + \cdots + a_n}{n}$。

2）构建 SparkContext 以及 SQLContext，导入 Spark 的 org.apache.spark.sql.functions._，将使用 spark sql 强大的内置函数功能。

3）构建 Row 类型的 Seq 集合变量 data，其中 1 到 1000 范围的数值，如数值小于 500，则返回 Row（A，数值）；如数值大于 500，则返回 Row（B，数值）；通过 sc 的 parallelize 方法读入 data 数据，生成 rdd。

4）构建 StructType 类型变量 schema，其元素为（（key，字符串类型），（value，浮点数类型））。将 rdd 和相应的 schema 通过 sqlContext.createDataFrame 方法构建 DataFrame df。

5）创建 CustomMean 实例 customMean。

6）在 Spark SQL 语句中使用实例化后的 CustomMean UDAF。df 的 DataFrame 根据 key 列分组，使用 agg 函数查询，查询结果是 2 列：

① 第 1 列：使用自定义的 CustomMean UDAF 计算算术平均数的函数 customMean，在 customMean 中传入 value 列数值，分别计算 Key 为 A 时，value 列小于 500 的算术平均数；计算 Key 为 B 时，value 列大于 500 的算术平均数。

② 第 2 列：使用 Spark SQL 内置的平均数函数 avg，在 avg 中传入 value 列数值，分别计算 Key 为 A 时，value 列小于 500 的算术平均数；计算 Key 为 B 时，value 列大于 500 的算术平均数。

自定义的 CustomMean UDAF 算术平均数函数计算结果和 Spark SQL 内置的平均数函数 avg 计算结果相同。

```
import org.apache.spark.{SparkContext,SparkConf}
import org.apache.spark.sql.{SQLContext,Row}
import org.apache.spark.sql.expressions.{MutableAggregationBuffer,UserDefinedAggregateFunction}
import org.apache.spark.sql.types._

objectmyUDAF{
    classCustomMean extends UserDefinedAggregateFunction{

        //输入数据类型为 Double
        def inputSchema:StructType = StructType(Array(StructField("item",DoubleType)))

        //数据的和为 sum，是 Double 类型，数据的个数为 cnt，是 Long 类型
        def bufferSchema = StructType(Array(StructField("sum",DoubleType),StructField("cnt",LongType)))

        //UDAF 的返回值为 Double 类型
        def dataType:DataType = DoubleType

        //在给定输入值的前提下，此 UDAF 总是生成一组相同的结果
```

第6章　Spark SQL UDF与UDAF

```scala
    def deterministic = true

    //sum 的初值为 0,cnt 的初值为 0
    def initialize(buffer:MutableAggregationBuffer) = {
      buffer(0) = 0.toDouble
      buffer(1) = 0L
    }
    //每输入一个数,将其与 sum 相加,同时 cnt 加 1
    def update(buffer:MutableAggregationBuffer,input:Row) = {
      buffer(0) = buffer.getDouble(0) + input.getDouble(0)
      buffer(1) = buffer.getLong(1) + 1
    }

    def merge(buffer1:MutableAggregationBuffer,buffer2:Row) = {
      buffer1(0) = buffer1.getDouble(0) + buffer2.getDouble(0)
      buffer1(1) = buffer1.getLong(1) + buffer2.getLong(1)
    }

    //最终的结果为:sum/cnt
    def evaluate(buffer:Row) = {buffer.getDouble(0)/buffer.getLong(1).toDouble}
  }

  def main(args:Array[String]) {
    val conf = newSparkConf().setAppName("MyUDAF").setMaster("local[*]")
    val sc = newSparkContext(conf)
    val sqlContext = new SQLContext(sc)

    import org.apache.spark.sql.functions._

    val data = (1 to 1000).map{x:Int => x match{
      case t if t<=500 => Row("A",t.toDouble)
      case t => Row("B",t.toDouble)
    }}
    val rdd = sc.parallelize(data)
    val schema = StructType(Array(StructField("key",StringType),StructField("value",DoubleType)))
    val df = sqlContext.createDataFrame(rdd,schema)

    //实例化 CustomMean,在 SQL 语句中直接使用实例化后的 UDAF
    val customMean = new CustomMean()
    df.groupBy("key").agg(customMean(df.col("value")).as("custom_mean"),avg("value").as("avg")).show()
  }
}
```

在本地运行，结果如下所示。

```
+---+-----------+-----+
|key|custom_mean| avg |
+---+-----------+-----+
| A |      250.5|250.5|
| B |      750.5|750.5|
+---+-----------+-----+
```

6.4.4　BelowThreshold 函数

函数名称：belowThreshold。

函数功能：用于检测分组的数据中是否存在位于给定阈值下的数，如果是，则返回 true；否，则返回 false。

函数示例：

1）定义 belowThreshold 类继承至 UserDefinedAggregateFunction：

- 重载实现方法 inputSchema：返回 StructType 字段（power，整数类型），作为 belowThreshold 函数的输入参数；
- 重载实现方法 BufferSchema：返回 StructType 字段（bool，布尔值类型），作为 belowThreshold 函数的中间结果的值；
- 重载实现方法 dataType：dataType 表示 belowThreshold 函数返回值的类型是布尔值类型；
- 重载实现方法 deterministic：设置 true，在给定输入值的前提下，belowThreshold 生成一组相同的结果；
- 重载实现方法 initialize：初始化 buffer 的第 0 个元素布尔值为 false，用于初始化聚集缓冲区（MutableAggregationBuffer）的值；
- 重载实现方法 update：如果读入 input 每行的第 0 个元素不为空，则将 buffer 第 0 个元素原布尔值与读入 input 每行的第 0 个元素 power 值是否小于"-40"进行逻辑或运算，两者之一为 true，更新 buffer 第 0 个元素值为 True。即读入数据中只要有一个小于"-40"的数据，就返回 true；
- 重载实现方法 merge：merge 用于合并两个聚集缓冲区，将第一个缓冲区的第 0 个元素的布尔值与第二个缓冲区的第 0 个元素的布尔值做逻辑或运算，作为合并以后的缓冲区的第 0 个元素的布尔值；
- 重载实现方法 evaluate：belowThreshold 函数的最终计算结果为：buffer 的第 0 个元素的布尔值。

2）构建 SparkContext 以及 SQLContext，导入 Spark 的 sqlContext 隐式转换类 import sqlContext.implicits._，用于将一个 RDD 隐式转换为一个 DataFrame。

3）通过 sc 的 parallelize 方法读入 Seq 集合数据，调用 toDF()方法转换成 DataFrame，df 包括两列：group，power。

第6章 Spark SQL UDF与UDAF

4）创建 belowThreshold 实例 belowThreshold。

5）在 sqlContext.udf 中注册 belowThreshold 的自定义函数，判断输入的数值是否小于"-40"。

6）df 的 DataFrame 根据 group 列分组，使用 agg 函数通过 belowThreshold UDAF 函数判断输入 power 列的数值是否小于"-40"，查询打印结果。

```scala
import org.apache.spark.{SparkContext,SparkConf}
import org.apache.spark.sql.{SQLContext,Row}
import org.apache.spark.sql.expressions.{MutableAggregationBuffer,
UserDefinedAggregateFunction}
import org.apache.spark.sql.types._

object myUDAF{
  class belowThreshold extends UserDefinedAggregateFunction{
    //输入类型为 Int 类型
    def inputSchema = new StructType().add("power",IntegerType)

    //中间类型为布尔值类型
    def bufferSchema = new StructType().add("bool",BooleanType)
    //返回类型为布尔值类型
    def dataType = BooleanType

    //在给定输入值的前提下,此 UDAF 总是生成一组相同的结果
    def deterministic = true

    //初始值为 false
    def initialize(buffer:MutableAggregationBuffer) = buffer.update(0,false)

    //注意两个 bool 值之间的"|"
    def update(buffer:MutableAggregationBuffer,input:Row) = {
      if(!input.isNullAt(0))
        buffer.update(0,buffer.getBoolean(0) | input.getInt(0) < -40)
    }

    def merge(buffer1:MutableAggregationBuffer,buffer2:Row) = {
      buffer1.update(0,buffer1.getBoolean(0) | buffer2.getBoolean(0))
    }

    def evaluate(buffer:Row) = buffer.getBoolean(0)
  }

  def main(args:Array[String]){
    val conf = newSparkConf().setAppName("MyUDAF").setMaster("local[*]")
```

```
        val sc = newSparkContext(conf)
        val sqlContext = new SQLContext(sc)

    import sqlContext.implicits._

        val df = sc.parallelize(Seq(("a",10),("a",20),("b",30),("b",-50))).toDF("group","power")

        //实例化 belowThreshold,向 Spark SQL 的功能注册表注册此 UDAF
        val belowThreshold = new belowThreshold()
sqlContext.udf.register("belowThreshold",belowThreshold)

        //在 Driver 上打印结果
        df.groupBy($"group").agg(belowThreshold($"power").alias("belowThreshold")).show

        sc.stop()
    }
}
```

在本地运行,结果如下所示:

```
+-----+--------------+
|group|belowThreshold|
+-----+--------------+
|    a|         false|
|    b|          true|
+-----+--------------+
```

6.4.5　YearCompare 函数

函数名称:YearCompare。

函数功能:统计今年同比去年销售金额的增长率。

本案例我们在 Spark-Shell 中操作 UDAF。

1) 首先,启动 Spark 集群,在命令行中输入 Spark-Shell,导入所需的类。

2) 定义一个日期范围的 case class DateRange,成员变量包括起始日期、终止日期;inMiddle 方法对传入的日期进行判断,如果日期在起始日期、终止日期之间,则返回 true 值。

3) 定义一个顾客信息的 case class Customer,成员变量包括 ID 号、名字、销售价格、折扣价格、州名、销售日期。

4) 定义 YearCompare 类继承至 UserDefinedAggregateFunction:

- YearCompare 私有方法 subtractOneYear:将传入的时间年份减1,即去年的时间日期;

第6章　Spark SQL UDF与UDAF

- YearCompare 私有成员变量 previous：将 YearCompare 传入的 current 日期的起始日期、终止日期分别减去 1 年，重新构建一个去年的 DateRange 日期范围，即（去年的起始日期、去年的终止日期）作为 previous；
- 重载实现方法 inputSchema：返回 StructType 字段（销售额 metric，浮点数类型），（时间日期，日期类型），作为 YearCompare 函数的输入参数；
- 重载实现方法 BufferSchema：返回 StructType 字段（今年销售额总量 sumOfCurrent，浮点数类型），（去年销售额总量 sumOfPrevious，浮点数类型）作为 YearCompare 函数的中间结果的值；
- 重载实现方法 dataType：dataType 表示 YearCompare 函数返回值的类型是浮点数类型；
- 重载实现方法 deterministic：设置 true，在给定输入值的前提下，YearCompare 生成一组相同的结果；
- 重载实现方法 initialize：初始化 buffer 的第 0 个元素今年销售额总量 sumOfCurrent 为 0.0，初始化 buffer 的第 1 个元素去年销售额总量 sumOfPrevious 为 0.0，用于初始化聚集缓冲区（MutableAggregationBuffer）的值；
- 重载实现方法 update：如果读入 input 行的第 1 个元素日期在 current 今年日期的起始日期、终止日期范围之内，则将 buffer 第 0 个元素今年销售额总量 sumOfCurrent 加上读入 input 行的第 0 个元素销售额 metric，即统计今年的销售额；如果读入 input 行的第 1 个元素日期在 previous 去年日期的起始日期、终止日期范围之内，则将 buffer 第 1 个元素去年销售额总量 sumOfPrevious 加上读入 input 行的第 0 个元素销售额 metric，即统计去年销售额总量；
- 重载实现方法 merge：merge 用于合并两个聚集缓冲区，将第一个缓冲区的第 0 个元素的今年销售额总量 sumOfCurrent 与第二个缓冲区的第 0 个元素的今年销售额总量 sumOfCurrent 相加，作为合并以后的缓冲区的第 0 个元素的今年销售额总量；将第一个缓冲区的第 1 个元素的去年销售额总量 sumOfPrevious 与第二个缓冲区的第 1 个元素的去年销售额总量 sumOfPrevious 相加，作为合并以后的缓冲区的第 1 个元素的去年销售额总量；
- 重载实现方法 evaluate：YearCompare 函数的最终计算结果为：如果缓冲区的第 1 个元素即去年销售额总量 sumOfPrevious 为 0，则返回 0；如果缓冲区的第 1 个元素的去年销售额总量 sumOfPrevious 不为 0，则将缓冲区的第 0 个元素即今年销售额总量 sumOfCurrent 减去缓冲区的第 1 个元素即去年销售额总量 sumOfPrevious，然后除以缓冲区的第 1 个元素即去年销售额总量 sumOfPrevious，然后乘以 100 计算出百分比，即今年同比去年销售金额的增长率。

5）导入 Spark 的 sqlContext 隐式转换类 import sqlContext.implicits._，用于将一个 RDD 隐式转换为一个 DataFrame。

6）构建 Seq 集合顾客信息的数据 data，包括 ID 号，名字，销售价格，折扣价格，州名，销售日期。

7）通过 sc 的 parallelize 方法读入 Seq 集合数据 data，调用 toDF（）方法转换成 DataFrame，通过 dataFrame.printSchema（）打印出 DataFrame 的结构。通过 registerTempTable（"salesInfo"）方法将 DataFrame 注册为临时表 salesInfo。

8）指定日期范围 current，实例化聚集函数 YearCompare，将日期范围 current 传入 YearCompare 函数。

9）向 Spark SQL 注册 yearCompare 的 UDAF 函数，从临时表 salesInfo 查询，调用 yearCompare 方法，在 yearCompare 方法中传入 salesInfo 的销售额、salesInfo 的日期，然后将传入的 salesInfo 日期与 current 日期、previous 日期进行判断，如果在今年的日期范围内，统计今年的销售额；如果在去年的日期范围内，统计去年的销售额，计算出今年同比去年销售金额的增长率。最后，打印输出结果。

```
scala > import java.sql.{Timestamp,Date}
import java.sql.{Timestamp,Date}

scala > import org.apache.spark.sql.Row
import org.apache.spark.sql.Row

scala > importorg.apache.spark.sql.expressions.{MutableAggregationBuffer, UserDefinedAggregate-
Function}
import org.apache.spark.sql.expressions.{MutableAggregationBuffer,UserDefinedAggregateFunction}

scala > import org.apache.spark.sql.types._
import org.apache.spark.sql.types._
```

创建 case class，如下所示：

```
scala > case classDateRange(startDate:Timestamp,endDate:Timestamp){
        |    //判断某一个日期是否处于特定的日期范围
        |    def inMiddle(targetDate:Date):Boolean = {
        |        targetDate.before(endDate)&&targetDate.after(startDate)
        |    }
        | }
defined classDateRange

scala > case class Customer(id:Integer,name:String,sales:Double,discounts:Double,state:String,sale-
Date:String)   //ID 号，名字，销售价格，折扣，州名，销售日期

defined class Customer
```

创建自定义的类 YearCompare，该类继承了 UserDefinedAggregateFunction：

```
scala >   classYearCompare(current:DateRange)extends UserDefinedAggregatcFunction{
        |    val previous:DateRange = DateRange(subtractOneYear(current.startDate),
        |                                       subtractOneYear(current.endDate))
```

```
//在输入的参数中,有 Double 类型的,也有 Date 类型的
def inputSchema:StructType = {
    StructType(StructField("metric",DoubleType)::StructField("timeCategory",DateType)::Nil)
}

//中间结果都是 Double 类型
def bufferSchema:StructType = {
        StructType(StructField("sumOfCurrent",DoubleType)::StructField("sumOfPrevious",DoubleType)::Nil)
}

//此 UDAF 的最终返回类型为 Double
def dataType:org.apache.spark.sql.types.DataType = DoubleType

//在给定输入值的前提下,此 UDAF 总是生成一组相同的结果
def deterministic:Boolean = true

def initialize(buffer:MutableAggregationBuffer):Unit = {
    buffer.update(0,0.0)
    buffer.update(1,0.0)
}

def update(buffer:MutableAggregationBuffer,input:Row):Unit = {
    if(current.inMiddle(input.getAs[Date](1))){
        buffer(0) = buffer.getAs[Double](0) + input.getAs[Double](0)
    }
    if(previous.inMiddle(input.getAs[Date](1))){
        buffer(1) = buffer.getAs[Double](0) + input.getAs[Double](0)
    }
}

def merge(buffer1:MutableAggregationBuffer,buffer2:Row):Unit = {
    buffer1(0) = buffer1.getAs[Double](0) + buffer2.getAs[Double](0)
    buffer1(1) = buffer1.getAs[Double](1) + buffer2.getAs[Double](1)
}
```

```
        |     def evaluate(buffer:Row):Any = {
        |       if(buffer.getDouble(1) == 0.0){
        |         0.0
        |       }else{
        |         (buffer.getDouble(0) - buffer.getDouble(1))/buffer.getDouble(1) * 100
        |       }
        |     }
        |
        |     //减掉一年
        |     private def subtractOneYear(date:Timestamp):Timestamp = {
        |       val prev = new Timestamp(date.getTime)
        |       prev.setYear(prev.getYear - 1)
        |       prev
        |     }
        |   }
defined class YearCompare
```

接下来，引入隐式转换：

```
scala > import sqlContext.implicits._     //引入隐式转换
import sqlContext.implicits._
```

输入数据，代码如下所示：

```
scala > val data = Seq(Customer(1,"Widget Co",120200.00,0.00,"AZ","2015-02-28"),
       |                Customer(2,"Acme Widgets",410600.00,560.00,"CA","2015-03-08"),
       |                Customer(3,"Widgetry",410550.00,230.00,"CA","2016-05-01"),
       |                Customer(4,"Widgets R Us",410505.00,0.0,"CA","2016-05-04"),
       |                Customer(5,"YeOldeWidgete",500.00,0.0,"MA","2016-08-15"))
data:Seq[Customer] = List(Customer(1,Widget Co,120200.0,0.0,AZ,2015-02-28),Customer
(2,Acme Widgets,410600.0,560.0,CA,2015-03-08),Customer(3,Widgetry,410550.0,230.0,
CA,2016-05-01),Customer(4,Widgets R Us,410505.0,0.0,CA,2016-05-04),Customer(5,Ye
OldeWidgete,500.0,0.0,MA,2016-08-15))
```

加载数据，使之成为 DataFrame，即数据框架：

```
scala > val dataFrame = sc.parallelize(data).toDF()
dataFrame:org.apache.spark.sql.DataFrame = [ID:int, Name:string, Sales:double, Discounts:double,
State:string, SaleDate:string]
//打印出 dataFrame 的 schema
scala > dataFrame.printSchema()
```

第6章 Spark SQL UDF与UDAF

```
root
 |-- ID: integer (nullable = true)
 |-- Name: string (nullable = true)
 |-- Sales: double (nullable = false)
 |-- Discounts: double (nullable = false)
 |-- State: string (nullable = true)
 |-- SaleDate: string (nullable = true)
```

```scala
//dataFrame 注册为表，表名为 salesInfo
scala> dataFrame.registerTempTable("salesInfo")

import java.sql.{Timestamp, Date}
import org.apache.spark.{SparkContext, SparkConf}
import org.apache.spark.sql.{SQLContext, Row}
import org.apache.spark.sql.expressions.{MutableAggregationBuffer, UserDefinedAggregateFunction}
import org.apache.spark.sql.expressions.{MutableAggregationBuffer, UserDefinedAggregateFunction}
import org.apache.spark.sql.types._

//指定日期范围
scala> val current = DateRange(Timestamp.valueOf("2016-01-01 00:00:00"), Timestamp.valueOf("2016-10-01 00:00:00"))
current: DateRange = DateRange(2016-01-01 00:00:00.0, 2016-10-01 00:00:00.0)

//实例化聚集函数
scala> val yearCompare = new YearCompare(current)
yearCompare: YearCompare = $iwC$$iwC$YearCompare@349c6b2d

//向 Spark SQL 的功能注册表注册名为 yearCompare 的 UDAF
scala> sqlContext.udf.register("yearCompare", yearCompare)
res1: org.apache.spark.sql.expressions.UserDefinedAggregateFunction = $iwC$$iwC$YearCompare@349c6b2d

//在 Driver 上打印出结果
scala> sqlContext.sql("select yearCompare(sales, saleDate) as yearCompare from salesInfo").show()
...
16/06/05 17:43:32 INFO scheduler.DAGScheduler: Job 0 finished: show at <console>:35, took 9.261529 s
+------------------+
|       yearCompare|
```

193

```
+------------------+
|54.77675207234364 |
+------------------+
```

6.4.6　WordCount 函数

函数名称：computeLength 及 UDAF 函数类 wordCount。

函数功能：computeLength 函数统计输入字符串的长度；UDAF 函数类 wordCount 统计输入单词的个数。

函数示例：这个示例在一条 SQL 语句中同时使用了 UDF 和 UDAF。其中，UDF 是用来计算字符串的长度，UDAF 则是完成词频统计的功能。

1）创建 SparkContext 对象及 SQLContext。

2）模拟实际使用的数据，构建数组 bigData。

3）基于提供的数据创建 DataFrame：通过 sc 的 parallelize 方法读入 bigData 数据；将 bigDataRDD 转换成 RDD［Row］；构建 StructType 变量（单词，字符串类型，允许为空）；通过 sqlContext.createDataFrame 方法根据 bigDataRDDRow，structType 构建 DataFrame，将 bigDataDF 注册为临时表 bigDataTable。

4）按照模板实现 UDAF：

- 重载实现方法 inputSchema：返回 StructType 字段（输入字符，字符串类型，允许为空），作为 MyUDAF 函数的输入参数；
- 重载实现方法 BufferSchema：返回 StructType 字段（计数次数，浮点类型，允许为空），作为 MyUDAF 函数的中间结果的值；
- 重载实现方法 dataType：dataType 表示 MyUDAF 函数返回值的类型是整型；
- 重载实现方法 deterministic：设置 true，在给定输入值的前提下，MyUDAF 生成一组相同的结果；
- 重载实现方法 initialize：初始化计数次数为 0，用于初始化聚集缓冲区（MutableAggregationBuffer）的值；
- 重载实现方法 update：每读入 1 行，将计数次数计为 1，在缓冲区的第 0 个元素原计数次数基础上累加 1 作为新的计数次数；
- 重载实现方法 merge：merge 用于合并两个聚集缓冲区，将第一个缓冲区的第 0 个元素计数值加上第二个缓冲区的第 0 个元素计数值，并将结果存储到 MutableAggregationBuffer；
- 重载实现方法 evaluate：MyUDAF 函数的最终计算结果为 buffer 的第 0 个元素（即单词累计计数次数）。

5）SQLContext UDF 函数的综合应用：计算字符串的长度。

- 在 sqlContext.udf 中注册 computeLength 的自定义函数：输入字符串，统计字符串的长度。
- 直接在 SQL 语句中使用 computeLength UDF，在临时表 bigDataTable 中查询，查询单词名、使用 computeLength 方法计算单词长度，打印输出结果。

第6章 Spark SQL UDF与UDAF

6）SQLContext UDAF 函数的综合应用：计算词频统计。
- 在 sqlContext.udf 中注册 wordCount 的自定义函数，将创建的 MyUDAF 实例传入。
- 在 sqlContext.sql 语句中使用 UDAF，在临时表 bigDataTable 中查询，根据单词分组，查询单词名、使用 MyUDAF UDAF 函数统计单词次数，使用 computeLength 方法计算单词长度，使用 show( ) 打印输出结果。

```scala
import org.apache.spark.sql.expressions.{MutableAggregationBuffer,
UserDefinedAggregateFunction}
import org.apache.spark.sql.types._
import org.apache.spark.sql.{Row,SQLContext}
import org.apache.spark.{SparkContext,SparkConf}

/**
 * 使用 Scala 开发集群运行的 SparkWordCount 程序
 * @author DT 大数据梦工厂
 * 新浪微博:http://weibo.com/ilovepains/
 * Created by hp on 2016/3/31.
 * 通过案例实战了解 Spark SQL 下的 UDF 和 UDAF 的具体使用:
 * 用户自定义的函数(User Defined Function,UDF),函数的输入是一条具体的数据记录,实际上
 就是普通的 Scala 函数;
 * 用户自定义的聚合函数(User Defined Aggregation Function,UDAF),函数本身作用于数据集
 合,能够在聚合操作的基础上进行自定义操作;
 *
 * 实际上,UDF 会被 Spark SQL 中的 Catalyst 封装成为 Expression,最终会通过 eval 方法来计算输
 入的数据 Row(此处的 Row 和 DataFrame 中的 Row 没有任何关系)
 *
 */
object SparkSQLUDFUDAF{

  def main(args:Array[String]){
    /**
      * 第1步:创建 Spark 的配置对象 SparkConf,设置 Spark 程序的运行时的配置信息,
      * 例如通过 setMaster 来设置程序要链接的 Spark 集群的 Master 的 URL,如果设置
      * 为 local,则代表 Spark 程序在本地运行,特别适合机器配置条件非常差(例如
      * 只有 1 GB 的内存)的初学者
      */
    val conf = newSparkConf()//创建 SparkConf 对象
    conf.setAppName("SparkSQLUDFUDAF")//设置应用程序的名称,在程序运行的监控界面可
以看到名称
    //conf.setMaster("spark://Master:7077")//此时,程序在 Spark 集群
    conf.setMaster("local[*]")

    /**
```

```
 * 第 2 步:创建 SparkContext 对象
 * SparkContext 是 Spark 程序所有功能的唯一入口,采用 Scala、Java、Python、R 等都必须有一
个 SparkContext
 * SparkContext 的核心作用:初始化 Spark 应用程序运行所需要的核心组件,包括 * DAG-
Scheduler、TaskScheduler、SchedulerBackend
 * 同时还会负责 Spark 程序往 Master 注册程序等
 * SparkContext 是整个 Spark 应用程序中最为至关重要的一个对象
 */
    val sc = newSparkContext(conf)//创建 SparkContext 对象,通过传入 SparkConf 实例来定制
Spark 运行的具体参数和配置信息

    val sqlContext = new SQLContext(sc)//构建 SQL 上下文
    //模拟实际使用的数据
    val bigData = Array("Spark","Spark","Hadoop","Spark","Hadoop","Spark","Spark","Ha-
doop","Spark","Hadoop")

    //基于提供的数据创建 DataFrame
    val bigDataRDD = sc.parallelize(bigData)
    val bigDataRDDRow = bigDataRDD.map(item => Row(item))
    val structType = StructType(Array(StructField("word",StringType,true)))
    val bigDataDF = sqlContext.createDataFrame(bigDataRDDRow,structType)

bigDataDF.registerTempTable("bigDataTable")//注册成为临时表

    /**
     * 通过 SQLContext 注册 UDF,在 Scala 2.10.x 版本中,UDF 函数最多可以接受 22 个输入
参数
     */
sqlContext.udf.register("computeLength",(input:String) => input.length)

    //直接在 SQL 语句中使用 UDF,就像使用 SQL 自动的内部函数一样
sqlContext.sql("select word,computeLength(word) as length from bigDataTable").show
    //注册 UDAF
sqlContext.udf.register("wordCount",new MyUDAF)
sqlContext.sql("select word,wordCount(word) as count,computeLength(word) as length from bigDataT-
able group by word").show()
    }
  }
/**
 * 按照模板实现 UDAF
 */
classMyUDAF extends UserDefinedAggregateFunction {
  /**
   * 该方法指定具体输入数据的类型
```

第6章　Spark SQL UDF与UDAF

```
  * @return
  */
override def inputSchema:StructType = StructType(Array(StructField("input",StringType,true)))

/**
  * 在进行聚合操作的时候所要处理的数据的结果类型
  * @return
  */
override def bufferSchema:StructType = StructType(Array(StructField("count",IntegerType,true)))

/**
  * 指定UDAF函数计算后返回的结果类型
  * @return
  */
override def dataType:DataType = IntegerType

override def deterministic:Boolean = true

/**
  * 在Aggregate之前每组数据的初始化结果
  * @param buffer
  */
override def initialize(buffer:MutableAggregationBuffer):Unit = {buffer(0) = 0}

/**
  * 在进行聚合的时候,每当有新的值进来,对分组后的聚合如何进行计算
  * 本地的聚合操作,相当于HadoopMapReduce模型中的Combiner
  * @param buffer
  * @param input
  */
override def update(buffer:MutableAggregationBuffer,input:Row):Unit = {
    buffer(0) = buffer.getAs[Int](0) + 1
}

/**
  * 最后在分布式结点进行Local Reduce完成后需要进行全局级别的Merge操作
  * @param buffer1
  * @param buffer2
  */
override def merge(buffer1:MutableAggregationBuffer,buffer2:Row):Unit = {
    buffer1(0) = buffer1.getAs[Int](0) + buffer2.getAs[Int](0)
}

/**
  * 返回UDAF最后的计算结果
```

```
* @param buffer
* @return
*/
override def evaluate(buffer:Row):Any = buffer.getAs[Int](0)
}
```

在本地运行，结果如下所示。

```
+------+------+
| word |length|
+------+------+
| Spark|    5 |
| Spark|    5 |
|Hadoop|    6 |
| Spark|    5 |
|Hadoop|    6 |
| Spark|    5 |
| Spark|    5 |
|Hadoop|    6 |
| Spark|    5 |
|Hadoop|    6 |
+------+------+

+------+-----+------+
| word |count|length|
+------+-----+------+
| Spark|   6 |    5 |
|Hadoop|   4 |    6 |
+------+-----+------+
```

Section 6.5 本章小结

  本章阐述了 Spark SQL UDF 与 UDAF 的使用，在实际生产环境中，数据库内置的函数并不一定能满足业务的需要，这时可以使用自定义的 UDF 构建能实现业务功能的函数，像内置的函数一样使用；UDAF 是用户自定义的聚合函数，函数本身作用于数据集合，能够在聚合操作的基础上进行自定义操作。

第 7 章　Thrift Server

Spark 从 1.1 开始增加了 CLI 和 Thrift Server，使得 Hive 用户很容易地上手 Spark SQL。Spark SQL 可以作为一个分布式查询引擎，使用 JDBC/ODBC 或命令行界面。在这种模式下，终端用户或应用程序可以直接与 Spark SQL 进行 SQL 查询交互，而无须编写任何代码。

本章将介绍 Thrift 的基本概念以及如何使用 Thrift Server 来与 Spark SQL 进行交互。

7.1　Thrift 概述

7.1.1　Thrift 的基本概念

Thrift 最初由 Facebook 用做系统内各语言之间的 RPC 通信。2007 年由 Facebook 贡献到 Apache 基金，之后发展成为一种可伸缩的、跨语言的服务开发框架。

📖 Apache Thrift——可伸缩的跨语言服务开发框架。

我们熟知的服务调用方式有很多种，比如：基于 SOAP 消息格式的 Web Service，基于 JSON 消息格式的 RESTful 等等。所使用的数据传输方式有：XML、JSON 等，然而 XML 相对体积太大，传输效率低，JSON 体积较小，但还不够完善。

Apache Thrift 采用接口描述语言定义并创建服务，支持可扩展的跨语言服务开发，所包含的代码生成引擎可以在多种语言中使用，如：Java、C++、Python、PHP、Ruby、Perl、C#等等。其传输数据采用二进制格式，相对 XML 和 JSON 体积更小，对于高并发、大数据量和多语言的环境更有优势。

Spark Thrift Server 是 Spark 框架中的一个应用程序，Spark Thrift Server 启动的时候，会启动 Spark Context、Spark SqlContext，最终调用第三方 Hive 框架中的 HiveServer2；而第三方 Hive 框架的 HiveServer2 不仅仅提供给 Spark Thrift Server 使用，而且作为 API 接口，可以直接提供 Java JDBC 连接 Hive Server2；第三方 Hive 框架中的 HiveServer2 使用 ApacheThrift 的协议开发，Thrift 是 Facebook 实现的一种高效的、支持多种编程语言的远程服务调用的框架。

Thrift 的基本概念包括数据类型、传输协议、传输层、服务端类型等。数据类型包含基本类型、结构体、容器、异常、服务类型；传输协议是 Thrift 客户端和服务器端远程调用传输数据采取的数据格式；传输层是数据传输方式；服务端类型包括单线程服务、多线程服务等。

199

1. 数据类型

Thrift 脚本可定义的数据类型包括以下几种类型：

1）基本类型：

- bool：布尔值，true 或 false，对应 Java 的 boolean。
- byte：8 位有符号整数，对应 Java 的 byte。
- i16：16 位有符号整数，对应 Java 的 short。
- i32：32 位有符号整数，对应 Java 的 int。
- i64：64 位有符号整数，对应 Java 的 long。
- double：64 位浮点数，对应 Java 的 double。
- string：未知编码文本或二进制字符串，对应 Java 的 String。

2）结构体类型：

- struct：定义公共的对象，类似于 C 语言中的结构体定义，在 Java 中是一个 JavaBean。

3）容器类型：

- list：对应 Java 的 ArrayList。
- set：对应 Java 的 HashSet。
- map：对应 Java 的 HashMap。

4）异常类型：

- exception：对应 Java 的 Exception。

5）服务类型：

- service：对应服务的类。

2. 协议

Thrift 可以让用户选择客户端与服务端之间传输通信协议的类别，在传输协议上总体划分为文本（Text）和二进制（Binary）传输协议，为节约带宽，提高传输效率，一般情况下使用二进制类型的传输协议为多数，有时也会使用基于文本类型的协议，这需要根据项目／产品中的实际需求。常用协议有以下几种：

1）TBinaryProtocol —— 二进制编码格式进行数据传输。

2）TCompactProtocol —— 高效率的、密集的二进制编码格式进行数据传输。

3）TJSONProtocol —— 使用 JSON 的数据编码协议进行数据传输。

4）TSimpleJSONProtocol —— 只提供 JSON 只写的协议，适用于通过脚本语言解析。

3. 传输层

Thrift 常用的传输层有以下几种：

1）TSocket —— 使用阻塞式 I/O 进行传输，是最常见的模式。

2）TFramedTransport —— 使用非阻塞方式，按块的大小进行传输，类似于 Java 中的 NIO。

3）TNonblockingTransport ——使用非阻塞方式，用于构建异步客户端。

4. 服务端类型

Thrift 常见的服务端类型有以下几种：

1）TSimpleServer —— 单线程服务器端使用标准的阻塞式 I/O。

2）TThreadPoolServer —— 多线程服务器端使用标准的阻塞式 I/O。

3）TNonblockingServer——多线程服务器端使用非阻塞式 I/O。

7.1.2　Thrift 的工作机制

我们在布置分布式大数据系统时，希望给数据分析师或者运营人员提供一种工具，他们可以通过 Web 方式提交 SQL 查询，或者说直接通过 Web 控制台来操作大数据系统。

要想达到这种目标，就需要一种特定的机制来访问我们的 Hadoop 或者 Spark。而 Thrift Server 正是起到这个作用，通过它就可以实现从 Web 的角度来访问 Spark SQL。图 7-1 所示是一个非常经典的 Spark SQL 企业应用实现架构，从中可以看出 Thrift Server 的作用：

JDBC/ODBC ➡ Thrift Server ➡ Spark SQL ➡ Hive 数据仓库

图 7-1　Spark SQL 企业应用实现架构

Thrift Server 在 JDBC/ODBC 和 Spark SQL 之间架设了一座桥梁；通过 JDBC/ODBC 接口访问 Thrift Server，Thrift Server 访问 Spark SQL。对用户而言，类似通过 JDBC/ODBC 访问 Oracle 数据库，用户通过 JDBC/ODBC 的接口访问 Thrift Server，相当于直接访问操作 Hive 数据仓库中的数据。

在实时性要求不高的情况下，JDBC/ODBC + Thrift Server + Spark SQL + Hive 数据仓库架构甚至可取代以传统关系数据库为后台的数据处理系统。

7.1.3　Thrift 的运行机制

Apache Thrift 包含一个完整的堆栈结构用于构建客户端和服务器端，Thrift 的体系架构如下图 7-2 所示：

Thrift 体系架构分析说明如下：

（1）用户开发者编写的脚本及代码

1）用户定义编写 Thrift 服务接口描述脚本，如 Hello.thrift。

2）用户编写服务器端业务代码：实现用户自己服务端的业务逻辑代码。

3）用户编写客户端业务代码：实现用户自己客户端的业务逻辑代码。

（2）Apache Thrift 框架自动编码

1）Apache Thrift 自动生成客户端框架代码。客户端框架代码按照服务接口描述文件生成。

2）Apache Thrift 根据用户定义的 Thrift 定义的服务接口描述文件（如 Hello.thrift）自动生成服务器端框架代码。

3）Apache Thrift 同时也生成数据的读写操作方法。

（3）使用 Thrift 选择不同的传输协议

例如：

使用 TCompactProtocol 协议构建的 HelloServiceServer.java；

TCompactProtocol.Factory proFactory = new TCompactProtocol.Factory();

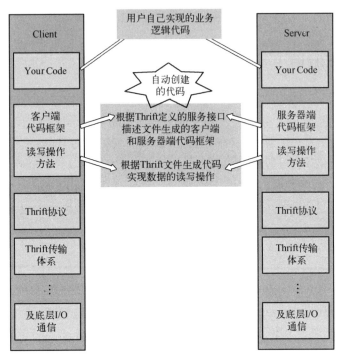

图 7-2 Thrift 体系架构图

使用 TCompactProtocol 协议的 HelloServiceClient.java：

```
TCompactProtocol protocol = new TCompactProtocol(transport);
```

（4）使用 Thrift 选择不同的传输层
例如：
使用 TFramedTransport 传输层构建的 HelloServiceServer.java：

```
// TNonblockingServerSocket 类继承 TNonblockingServerTransport
TNonblockingServerTransport serverTransport;
serverTransport = new TNonblockingServerSocket(10005);
Hello.Processor processor = new Hello.Processor(new HelloServiceImpl());
TServer server = new TNonblockingServer(processor, serverTransport);
System.out.println("Start server on port 10005 ...");
server.serve();
```

使用 TFramedTransport 传输层的 HelloServiceClient.java：

```
TTransport transport = new TFramedTransport(new TSocket("localhost", 10005));
```

（5）Thrift 支持的服务模型：
- TSimpleServer——简单的单线程服务模型，常用于测试。
- TThreadedServer——多线程服务模型，使用阻塞式 IO，每个请求创建一个线程。

第7章 Thrift Server

- TThreadPoolServer——线程池服务模型,使用标准的阻塞式 IO,预先创建一组线程处理请求。
- TNonblockingServer——多线程服务模型,使用非阻塞式 IO(需使用 TFramedTransport 数据传输方式)。

使用 TSimpleServer 服务端构建的 HelloServiceServer.java

```
TServerSocket serverTransport = new TServerSocket(7911);
TProcessor processor = new Hello.Processor(new HelloServiceImpl());
TServer server = new TSimpleServer(processor,serverTransport);
System.out.println("Start server on port 7911...");
server.serve();
```

7.1.4 一个简单的 Thrift 实例

本节讲解的 Thrift 实例,是 Thrift 开发的"Hello World"示例。整个过程为:先启动 Thrift 服务端服务,再启动客户端,客户端调用服务端 helloVoid 方法,实现在服务端打印输出"Hello World"。本节 Thrift 例子包括服务器代码编写、客户端代码编写、Thrift 服务接口描述脚本编写。在7.4节还将讲解 Spark Thrift Server 应用示例,通过 JDBC 访问 Spark Thrift Server,Spark Thrift Server 访问 Spark SQL,然后通过 Spark SQL 操作 Hive 数据库的数据。

本节 Thrift 例子中服务器编写的一般步骤如下:

1)创建 Handler:消息处理者,负责消息的发送及处理。
2)基于 Handler 创建 Processor。
3)创建 Transport。
4)创建 Protocol 方式。
5)基于 Processor,Transport 和 Protocol 创建 Server。
6)运行 Server。

客户端编写的一般步骤如下:

1)创建 Transport。
2)创建 Protocol 方式。
3)基于 Transport 和 Protocol 创建 Client。
4)运行 Client 的方法。

我们将通过这个简单的 Thrift 实现示例,来让大家直观地了解什么是 Thrift 以及如何使用 Thrift 构建服务。

下面将创建一个简单的服务 Hello,实现 helloString、helloInt、helloBoolean、helloVoid、helloNull 等接口功能。可以在集成开发环境如 IntelliJ IDEA Community Edition、Eclipse Jee Neon 进行开发。Hello 业务逻辑的服务端、客户端、Thrift 的语法规范文件由用户自己开发编写,Thrift 框架的服务器、客户端框架代码由 Apache Thrift 框架自动编码。

1)首先根据 Thrift 的语法规范编写脚本文件:Hello.thrift。Hello.thrift 可以使用记事本编辑,也可以使用集成开发环境如 IntelliJ IDEA Community Edition、Eclipse Jee Neon 等开发工具进行编辑,文件后缀名使用 .thrift 保存。

```thrift
namespace java service.demo
    service Hello{
        string helloString(1:string para)
        i32 helloInt(1:i32 para)
        bool helloBoolean(1:bool para)
        void helloVoid()
        string helloNull()
    }
```

其中定义了服务 Hello 的五个方法,每个方法包含一个方法名,参数列表和返回类型。每个参数包括参数序号,参数类型以及参数名。

Thrift 是对 IDL(Interface Definition Language)描述性语言的一种具体实现。因此,以上的服务描述文件使用 IDL 语法编写。使用 Thrift 工具编译 Hello.thrift,就会生成相应的 Hello.java 文件。该文件包含了在 Hello.thrift 文件中描述的服务 Hello 的接口定义,即 Hello.Iface 接口,以及服务调用的底层通信细节,包括客户端的调用逻辑 Hello.Client 以及服务器端的处理逻辑 Hello.Processor,用于构建客户端和服务器端的功能。

2)创建 HelloServiceImpl.java 文件并实现 Hello.java 文件中的 Hello.Iface 接口。

```java
package service.demo;
import org.apache.thrift.TException;
public class HelloServiceImpl implements Hello.Iface{
    @Override        //如果参数是 boolean 类型,返回 boolean 类型值
    public boolean helloBoolean(boolean para) throws TException{
        return para;
    }
    @Override        //如果参数是 int 类型,返回 int 类型值
    public int helloInt(int para) throws TException{
        try{
            Thread.sleep(20000);
        }catch(InterruptedException e){
            e.printStackTrace();
        }
        return para;
    }
    @Override        //如果参数是 null,返回 null 的字符串
    public String helloNull() throws TException{
        return null;
    }
    @Override        //如果参数是 String 类型,返回字符串值
    public String helloString(String para) throws TException{
        return para;
    }
```

```
        @Override           //如果参数为空,打印输出 Hello World
        public voidhelloVoid() throws TException{
            System.out.println("Hello World");
        }
    }
```

3）创建服务器端实现代码文件：HelloServiceServer.java，将 HelloServiceImpl 作为具体的处理器传递给 Thrift 服务器。

```
package service.server;
import org.apache.thrift.TProcessor;
import org.apache.thrift.protocol.TBinaryProtocol;
import org.apache.thrift.protocol.TBinaryProtocol.Factory;
import org.apache.thrift.server.TServer;
import org.apache.thrift.server.TThreadPoolServer;
import org.apache.thrift.transport.TServerSocket;
import org.apache.thrift.transport.TTransportException;
import service.demo.Hello;
import service.demo.HelloServiceImpl;

public class HelloServiceServer{
    /**
     * 启动 Thrift 服务器
     * @param args
     */
    public static void main(String[] args){
        try{
            //设置服务端口为 7911
            TServerSocketserver Transport = new TServerSocket(7911);
            //设置协议工厂为 TBinaryProtocol.Factory
            FactoryproFactory = new TBinaryProtocol.Factory();
            //关联处理器与 Hello 服务的实现
            TProcessor processor = new Hello.Processor(new HelloServiceImpl());
            TServer server = new TThreadPoolServer(processor,serverTransport,
    proFactory);
            System.out.println("Start server on port 7911...");
            server.serve();
        }catch(TTransportException e){
            e.printStackTrace();
        }
    }
}
```

4）创建客户端实现代码文件：HelloServiceClient.java，调用 Hello.client 访问服务端的逻辑实现。

```java
package service.client;
import org.apache.thrift.TException;
import org.apache.thrift.protocol.TBinaryProtocol;
import org.apache.thrift.protocol.TProtocol;
import org.apache.thrift.transport.TSocket;
import org.apache.thrift.transport.TTransport;
import org.apache.thrift.transport.TTransportException;
import service.demo.Hello;

public class HelloServiceClient{
    /**
     * 调用 Hello 服务
     * @param args
     */
    public static void main(String[] args){
        try{
            //设置调用的服务地址为本地,端口为 7911
            TTransport transport = new TSocket("localhost",7911);
            transport.open();
            //设置传输协议为 TBinaryProtocol
            TProtocol protocol = new TBinaryProtocol(transport);
            Hello.Client client = new Hello.Client(protocol);
            //调用服务的 helloVoid 方法
            client.helloVoid();
            transport.close();
        }catch(TTransportException e){
            e.printStackTrace();
        }catch(TException e){
            e.printStackTrace();
        }
    }
}
```

代码编写完后，先运行服务器 HelloServiceServer.java，再启动客户端 HelloServiceClient.java 调用服务 Hello 的方法 helloVoid，可在服务器端的控制台窗口输出"Hello World"。

7.2 Thrift Server 的启动过程

我们将通过 7.2.1 和 7.2.2 两节的内容，介绍 Thrift Sever 的启动过程，以及解析 HiveThriftServer2 类的实现。

第7章 Thrift Server

在 1.2.1 版本的 Hive 中，通过 HiveServer2 实现了 Thrift JDBC/ODBC 服务。可以通过 beeline 命令终端来测试 Thrift JDBC 服务。

HiveServer2（HS2）是一个服务器的接口，使远程客户端通过 Hive 来执行查询和检索结果。HiveServer2 基于 Thrift RPC，是 HiveServer 的一个改进版，支持多用户并发 HiveServer 和认证。它的目的是提供开放的 API 接口给 JDBC 和 ODBC 更好的支持。

7.2.1 Thrift Sever 启动详解

Thrift Server 启动的时候，其实是启动了一个 Spark SQL 的应用程序，同时开启一个侦听器，等待 JDBC 客户端的连接和提交查询。所以在配置 Thrift Server 的时候，可以在配置文件中配置 Thrift Server 的主机名和端口，如果要使用 Hive 数据的话，还要提供 Hive Metastore 的 uris。通常，可以在 conf/hive-site.xml 中定义以下几项配置，也可以使用环境变量进行配置（环境变量的优先级高于 hive-site.xml）。

我们查看一下本示例的 hive-site.xml 配置：

```
root@Master:/usr/local/spark/spark-1.6.3-bin-hadoop2.6/conf# cat hive-site.xml
<configuration>
<property>
<name>hive.metastore.uris</name>
<value>thrift://Master:9083</value>
<description>Thrift URI for the remotemetastore. Used by metastore client to connect to remote metastore.</description>
</property>
</configuration>
```

在当前 hive-site.xml 中只配置了 Hive Metastore 的 uris 的信息，其他配置采用默认配置。
接下来详细讲解 Thrift Sever 的启动过程。为了让大家深刻理解 Thrift Server 的启动过程，我们这里"挖一个坑"：

先把 HDFS 和 Spark 集群启动起来，但是此时先不启动 Hive Metastore 服务。为了稍后方便通过 Spark Web 控制台来查看 Job 的运行日志，此时需要把 HistoryServer 启动起来。

- 在 Hadoop 目录下启动 start-dfs.sh 和 start-yarn.sh，执行命令：

```
root@Master:/usr/local/hadoop/hadoop-2.7.1/sbin# ./start-all.sh
```

- 在 Spark 目录下执行命令：

```
root@Master:/usr/local/spark/spark-1.6.3-bin-hadoop2.6/sbin# ./start-all.sh
```

- 在 Spark 目录下启动 HistoryServer：

```
root@Master:/usr/local/spark/spark-1.6.3-bin-hadoop2.6/sbin# ./start-history-server.sh
```

- 通过 jps 查看启动的进程，至少包含下面的进程：

```
root@ Master:~# jps
    3426 Master
    2964 SecondaryNameNode
    3525 HistoryServer
    109654 Jps
    2733 NameNode
    3133 ResourceManager
```

📖 此时我们特意没有开启 Hive Metastore 元数据服务。

1. 尝试启动 Thrift Server

为了启动 Thrift Server，请在 Spark 目录运行下面的命令：

```
root@ Master:/usr/local/spark/spark-1.6.3-bin-hadoop2.6#./sbin/start-thriftserver.sh --master
spark://master:7077
starting org.apache.spark.sql.hive.thriftserver.HiveThriftServer2,logging to
/usr/local/spark/spark-1.6.3-bin-hadoop2.6/logs/spark-root-org.apache.spark.sql.hive.
thriftserver.HiveThriftServer2-1-Master.out
```

在命令终端上将会显示提示：正在启动 HiveThriftServer2，并记录到对应的日志文件。此时，我们可以通过 VIM 编辑器来打开该日志文件，命令如下：

```
root@ Master:/usr/local/spark/spark-1.6.3-bin-hadoop2.6# vim
/usr/local/spark/spark-1.6.3-bin-hadoop2.6/logs/spark-root-org.apache.spark.sql.hive.thrift-
server.HiveThriftServer2-1-Master.out
```

可以发现日志中有异常信息。在 74 行处（以实际为准），可以看见有异常信息：

```
73 16/04/30 17:04:55 INFO hive.metastore:Waiting 1 seconds before next connection attempt.
74 16/04/30 17:04:56 WARN metadata.Hive:Failed to accessmetastore. This class should not accessed
   in runtime.
75  org.apache.hadoop.hive.ql.metadata.HiveException:java.lang.RuntimeException:Unable to in-
stantiate org.apache.hadoop.hive.ql.metadata.SessionHiveMetaStoreClient
76          at org.apache.hadoop.hive.ql.metadata.Hive.getAllDatabases(Hive.java:1236)
77          at org.apache.hadoop.hive.ql.metadata.Hive.reloadFunctions(Hive.java:174)
```

通过上面的日志信息，可以发现因为 Hive 的 Metastore 服务没有启动，所以导致访问 Metastore 失败。此处就是为了让大家要注意分析日志文件信息，Spark 的很多日志文件信息，包括通过 Web 控制台都可以用来理解 Spark 的内部运行机制。

2. 正常启动 Thrift Server（以默认传输通道模式：binary 模式）

1）首先，正常启动 Hive Metastore 服务，命令如下：

```
root@ Master:/usr/local/spark/spark-1.6.3-bin-hadoop2.6# hive --service metastore&
```

或者也可以采用添加输出日志文件的方式启动：

```
hive --servicemetastore > Metastore.log 2>&1&
```

此时，通过 ps 命令查看 Hive Metastore 进程：

```
root@Master:/usr/local/spark/spark-1.6.3-bin-hadoop2.6# ps -ef | grep hive
root 5972 2559 22 17:35 pts/12   00:00:09/usr/lib/java/jdk1.8.0_66/bin/java -Xmx256m
-Djava.net.preferIPv4Stack=true -Dhadoop.log.dir=/usr/local/hadoop/hadoop-2.7.1/logs
-Dhadoop.log.file=hadoop.log -Dhadoop.home.dir=/usr/local/hadoop/hadoop-2.7.1 -Dhadoop.id.str=root
-Dhadoop.root.logger=INFO,console -Djava.library.path=/usr/local/hadoop/hadoop-2.7.1/lib/native
-Dhadoop.policy.file=hadoop-policy.xml -Djava.net.preferIPv4Stack=true -Xmx512m
-Dhadoop.security.logger=INFO,NullAppender org.apache.hadoop.util.RunJar
/usr/local/hive/apache-hive-1.2.1-bin/lib/hive-service-1.2.1.jar
org.apache.hadoop.hive.metastore.HiveMetaStore
```

可以发现 HiveMetaStore 服务已经正确启动。

📖 Metastore 是 Hive 元数据的集中存放地，元数据包括表的名字、表的列和分区及其属性等，通常存储在关系数据库如 MySQL、Derby 中。

2) 此时再次运行 start-thriftserver.sh，执行命令如下：

```
root@Master:/usr/local/spark/spark-1.6.3-bin-hadoop2.6# ./sbin/start-thriftserver.sh --master spark://master:7077
```

打开 spark-root-org.apache.spark.sql.hive.thriftserver.HiveThriftServer2-1-Master.out 日志文件，没有发现异常信息。我们仔细阅读日志文件内容，尤其是下面这段日志内容很关键：

```
Spark Command:/usr/lib/java/jdk1.8.0_66/bin/java -cp
/usr/local/spark/spark-1.6.3-bin-hadoop2.6/conf/:/usr/local/spark/spark-1.6.3-bin-hadoop2.6/lib/spark-assembly-1.6.0-hadoop2.6.0.jar:/usr/local/spark/spark-1.6.3-bin-hadoop2.6/lib/datanucleus-api-jdo-3.2.6.jar:/usr/local/spark/spark-1.6.3-bin-hadoop2.6/lib/datanucleus-core-3.2.10.jar:/usr/local/spark/spark-1.6.3-bin-hadoop2.6/lib/datanucleus-rdbms-3.2.9.jar:/usr/local/hadoop/hadoop-2.7.1/etc/hadoop/ -Xms2G -Xmx2G
org.apache.spark.deploy.SparkSubmit --master spark://master:7077 --class
org.apache.spark.sql.hive.thriftserver.HiveThriftServer2 spark-internal
========================================
16/05/11 11:52:06 INFOthriftserver.HiveThriftServer2:Starting SparkContext
```

通过上面这段日志信息，我们可以获知：启动 Spark SQL Thrift Server，其实就是调用 SparkSubmit 向服务器集群提交一个 class 为 HiveThriftServer2 的普通 Spark 应用程序。

当然，我们直接使用 Linux 的 ps 命令也可以查看 Thrift Server 具体的运行信息，如下所示：

```
root@Master:/usr/local/spark/spark-1.6.3-bin-hadoop2.6/conf# ps -ef | grep thriftserver
root 125080 1 1 11:52 pts/12 00:03:13 /usr/lib/java/jdk1.8.0_66/bin/java -cp
/usr/local/spark/spark-1.6.3-bin-hadoop2.6/conf/:/usr/local/spark/spark-1.6.3-bin-ha
doop2.6/lib/spark-assembly-1.6.0-hadoop2.6.0.jar:/usr/local/spark/spark-1.6.3-bin-ha
doop2.6/lib/datanucleus-api-jdo-3.2.6.jar:/usr/local/spark/spark-1.6.3-bin-hadoop2.6/
lib/datanucleus-core-3.2.10.jar:/usr/local/spark/spark-1.6.3-bin-hadoop2.6/lib/datanucle
us-rdbms-3.2.9.jar:/usr/local/hadoop/hadoop-2.7.1/etc/hadoop/ -Xms2G -Xmx2G org.a
pache.spark.deploy.SparkSubmit --master spark://master:7077 --class org.apache.spark.sql.
hive.thriftserver.HiveThriftServer2 spark-internal
```

  📖 Thrift Server 其实就是一个 class 为 HiveThriftServer2 的 Spark 应用程序。

3. 以 HTTP 传输通道模式启动 Thrift Server

  Thrift Server 的传输通道模式可以是 Binary 或者 HTTP，通过 HTTP 传输通道，Spark SQL Thrift Server 也支持发送 Thrift RPC 消息，这种方式在客户端和服务器之间支持代理中介是特别有用的（例如：负载均衡和安全原因）。

  使用下面的配置以支持 HTTP 模式，如表 7-1 所示。

表 7-1 使用 HTTP 模式的参数表

设 置 参 数	默 认	描 述
hive.server2.transport.mode	binary	值为 http，允许以 HTTP 通道模式传输
hive.server2.thrift.http.port	10001	HTTP 端口
hive.server2.thrift.http.path	cliservice	服务端点名称

  以上参数可以在 JDBC Connection URLs 系统参数中传入或者直接配置在 Hive 配置信息文件 hive-site.xml 中。

  1）先停止刚才已经启动的 Thrift Sever：

```
root@Master:/usr/local/spark/spark-1.6.3-bin-hadoop2.6/sbin# ./stop-thriftserver.sh
stopping org.apache.spark.sql.hive.thriftserver.HiveThriftServer2
```

  2）指定以 HTTP 模式再次运行 start-thriftserver.sh，执行命令如下：

```
root@Master:/usr/local/spark/spark-1.6.3-bin-hadoop2.6# ./start-thriftserver.sh --master
spark://master:7077 --hiveconf hive.server2.transport.mode=http --hiveconf hive.server2.
thrift.http.path=cliservice
```

  此时 HTTP 端口没有指定，默认值为 10001，正如表 7-1 所示。
  现在我们再次使用 ps 命令查看 Thrift Server 具体的运行信息，如下所示：

```
root@Master:/usr/local/spark/spark-1.6.3-bin-hadoop2.6/sbin# ps -ef | grep thriftserver
root 131000 1 99 17:25 pts/14 00:00:15 /usr/lib/java/jdk1.8.0_66/bin/java -cp
/usr/local/spark/spark-1.6.3-bin-hadoop2.6/conf/:/usr/local/spark/spark-1.6.3-bin-
```

第7章 Thrift Server

hadoop2.6/lib/spark - assembly - 1.6.0 - hadoop2.6.0.jar:/usr/local/spark/spark - 1.6.3 - bin - hadoop2.6/lib/datanucleus - api - jdo - 3.2.6.jar:/usr/local/spark/spark - 1.6.3 - bin - hadoop2.6/lib/datanucleus - core - 3.2.10.jar:/usr/local/spark/spark - 1.6.3 - bin - hadoop2.6/lib/datanucleus - rdbms - 3.2.9.jar:/usr/local/hadoop/hadoop - 2.7.1/etc/hadoop/ - Xms2G - Xmx2G org.apache.spark.deploy.SparkSubmit -- master spark://master:7077 -- class org.apache.spark.sql.hive.thriftserver.HiveThriftServer2 spark - internal -- hiveconf hive.server2.transport.mode = http -- hiveconf hive.server2.thrift.http.path = cliservice

3）此时用 netstat 命令可以查看 10001 端口已经处于侦听状态：

root@ Master:/usr/local/spark/spark - 1.6.3 - bin - hadoop2.6/conf# netstat - aptl | grep 10001
tcp6 0 0 [::]:10001 [::]:* LISTEN 1956/java

当 Thrift Server 以 HTTP 模式启动后，就可以被 beeling 或者 JDBC 客户端程序通过 HTTP 通道模式来进行连接操作了。

4. 查看 ThriftServer 命令参数

运行下面的命令：

root@ Master:/usr/local/spark/spark - 1.6.3 - bin - hadoop2.6/sbin# ./start - thriftserver.sh -- help

通过使用 -- help 参数，可以列出 ThriftServer 的命令参数，如表 7-2 所示：

Usage:./sbin/start - thriftserver[options][thrift server options]

表 7-2 ThriftServer 命令参数

参 数 名 称	功　　能
-- master MASTER_URL	指定 Master 位置。例如：spark://host:port,mesos://host:port,yarn,或者 Local
-- deploy - mode DEPLOY_MODE	以客户端模式启动 Driver（"client"），或以集群模式启动（"cluster"）。默认以 Client 模式启动
-- class CLASS_NAME	应用程序的主类（Java/Scala 程序）
-- name NAME	应用程序名称
-- jars JARS	以逗号分隔的本地 jar 包列表，将包含在 Driver 和 Executor 节点的 classpaths 中
-- packages	以逗号分隔的 maven 坐标，jar 包将包含在 Driver 和 Executor 节点的 classpaths 中，将自动搜索本地 maven 仓库、maven 中央仓库或者以 -- repositories 指定的其它远程仓库。坐标的格式为：groupId:artifactId:version
-- exclude - packages	以逗号分隔的 "groupId:artifactId" 列表，排除一些依赖，用于解决依赖冲突
-- repositories	以逗号分隔的远程仓库列表，用于查找给定的 -- packages 的 maven 坐标
-- py - files PY_FILES	以逗号分隔的 zip、egg 或 .py 文件，放在 Python 应用程序的路径 PYTHONPATH
-- files FILES	以逗号分隔的文件列表，将要放在 Executor 节点中的目录
-- conf PROP = VALUE	任意 Spark 配置属性
-- properties - file FILE	加载额外属性的文件路径。如果没有指定，将查找 conf/spark - defaults.conf 配置文件
-- driver - memory MEM	设定 Driver 内存（例如 1000 MB，2 GB）（默认 512 MB）
-- driver - java - options	设定 Driver 额外的 java 选项

211

（续）

参数名称	功能
-- driver - library - path	设定 Driver 额外的库路径
-- driver - class - path	设定 Driver 额外的类路径。注意：使用 -- jars 增加 jars 包会自动包含在类路径 classpath
-- executor - memory MEM	设定每个 Executor 的内存（例如：1000 MB，2 GB）（默认 1 GB）
-- help，- h	显示此帮助消息并退出
-- verbose，- v	打印额外的 debug 输出

以上［options］参数列表，完全与 spark - submit 应用程序的参数一样，如果不设置 master 参数，Spark 应用程序将在启动 Thrift Server 的机器中以 local 方式运行，可以通过 http：//机器名：4040 进行监控。如果在集群中运行 Thrift Server，一定要配置 master、executor - memory 等参数。再通过 hive - conf 来指定［thrift server options］系列参数，因为参数比较多，通常使用 conf/hive - site. xml 进行配置。

5. 退出 Thrift Server

Thrift Server 启动后处于监听状态，可以执行下面命令退出 ThriftServer：

```
root@ Master:/usr/local/spark/spark - 1. 6. 3 - bin - hadoop2. 6/sbin# ./stop - thriftserver. sh
stopping org. apache. spark. sql. hive. thriftserver. HiveThriftServer2
```

7.2.2　HiveThriftServer2 类的解析

通过 7.2.1 一节的学习，我们输入 start - thriftserver. sh 命令，Spark SQL Thrift Server 启动时调用 org. apache. spark. deploy. SparkSubmit，SparkSubmit 向服务器集群提交一个 class 为 HiveThriftServer2 的应用程序，HiveThriftServer2 类是一个 Spark 应用程序，也是 main 方法的入口类，因此，很有必要对其进行解析。

下面是 HiveThriftServer2 的 UML 类图，如图 7-3 所示。

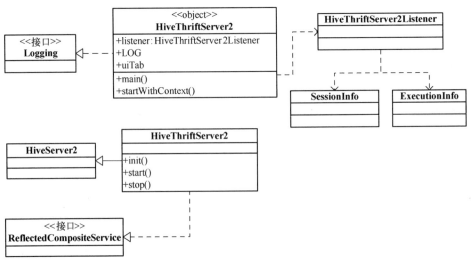

图 7-3　HiveThriftServer2UML 类图

第7章 Thrift Server

伴生对象 object HiveThriftServer2 继承至 Logging，是 Spark SQL 中连接 HiveServer2 的主入口点，启动 SparkSQLContext 和 HiveThriftServer2thrift server 服务。伴生类 class HiveThriftServer2 继承至 HiveServer2 及 ReflectedCompositeService，其中 HiveServer2 是 Hive 包中的类 org.apache.hive.service.server.{HiveServerServerOptionsProcessor, HiveServer2}。

下面是主要的几个关键点。

1. init 方法

```
override def init(hiveConf:HiveConf) {
    val sparkSqlCliService = new SparkSQLCLIService(this, hiveContext)
    set SuperField(this, "cliService", sparkSqlCliService)
    add Service(sparkSqlCliService)

    val thriftCliService = if(isHTTPTransportMode(hiveConf)) {
        new ThriftHttpCLIService(sparkSqlCliService)
    } else {
        new ThriftBinaryCLIService(sparkSqlCliService)
    }
    set SuperField(this, "thriftCLIService", thriftCliService)
    add Service(thriftCliService)
    initCompositeService(hiveConf)
}
```

HiveThriftServer2 继承了 Hive 的 HiveServer2 和 ReflectedCompositeService 两个类，并且复写了 init 方法。在初始化（HiveConf）的时候，会添加 cliService 和 thriftCLIService 两个服务，作为后台守护进程。其中，cliService 服务是 SparkSQLCLIService 对象的实例；thriftCLIService 服务在实例化时，会根据 HiveConf 参数进行判断，如果传输模式是 HTTP，就使用 ThriftHttpCLIService 对象，否则使用 ThriftBinaryCLIService 对象。ThriftHttpCLIService 和 ThriftBinaryCLIService 都是 ThriftCLIService 的子类，无论是哪个 ThriftCLIService 对象，都传入了 SparkSQLCLIService 的引用，Thrift 只是一个封装。

在添加这两个服务之后，再调用 initCompositeService 方法，把所有的服务启动起来。

2. 监听器

在 HiveThriftServer2 伴生对象中，有一个 listener 对象作为监听器，它是 HiveThriftServer2Listener 类的实例。listener 用来接收远程的请求，向 SparkContext 注册该监听器，把接收到的请求交给 HiveThriftServer2 类，再通过 Spark SQL 去执行任务。

3. main 方法入口

在 HiveThriftServer2 伴生对象中，还有 main 函数入口。正如 Thrift Server 启动日志所见：

```
Spark Command:/usr/lib/java/jdk1.8.0_66/bin/java – cp
/usr/local/spark/spark – 1.6.3 – bin – hadoop2.6/conf/:/usr/local/spark/spark – 1.6.3 – bin – hadoop2.6/lib/spark – assembly – 1.6.0 – hadoop2.6.0.jar:/usr/local/spark/spark – 1.6.3 – bin – hadoop2.6/lib/datanucleus – api – jdo – 3.2.6.jar:/usr/local/spark/spark – 1.6.3 – bin – hadoop2.6/lib/
```

```
datanucleus-core-3.2.10.jar:/usr/local/spark/spark-1.6.3-bin-hadoop2.6/lib/datanucleus-
rdbms-3.2.9.jar:/usr/local/hadoop/hadoop-2.7.1/etc/hadoop/ -Xms2G -Xmx2G
org.apache.spark.deploy.SparkSubmit --master spark://master:7077 --class
org.apache.spark.sql.hive.thriftserver.HiveThriftServer2 spark-internal
```

从 Spark Command 可以发现，启动 Thrift Sever 时，实际是调用 SparkSubmit 来提交 HiveThriftServer2 类的，所以该类必须要有 main 方法入口，用来创建一个新的进程。

在 main 方法中，主要处理逻辑是：创建 HiveThriftServer2 实例，以及创建 HiveThriftServer2Listener 监听器。具体代码如下：

```
def main(args: Array[String]) {
    val optionsProcessor = new HiveServerServerOptionsProcessor("HiveThriftServer2")
    if(!optionsProcessor.process(args)) {
      System.exit(-1)
    }

    logInfo("Starting SparkContext")
    SparkSQLEnv.init()
    ShutdownHookManager.addShutdownHook { () =>
      SparkSQLEnv.stop()
      uiTab.foreach(_.detach())
    }

    try {
      val server = new HiveThriftServer2(SparkSQLEnv.hiveContext)
      server.init(SparkSQLEnv.hiveContext.hiveconf)
      server.start()
      logInfo("HiveThriftServer2 started")
      listener = new HiveThriftServer2Listener(server, SparkSQLEnv.hiveContext.conf)
      SparkSQLEnv.sparkContext.addSparkListener(listener)
      uiTab = if(SparkSQLEnv.sparkContext.getConf.getBoolean("spark.ui.enabled", true)) {
          Some(new ThriftServerTab(SparkSQLEnv.sparkContext))
        } else {
          None
        }
      //If application was killed before HiveThriftServer2 start successfully then SparkSubmit
      //process can not exit, so check whether ifSparkContext was stopped.
      if(SparkSQLEnv.sparkContext.stopped.get()) {
        logError("SparkContext has stopped even if HiveServer2 has started, so exit")
        System.exit(-1)
      }
    } catch {
```

第7章 Thrift Server

```
        case e:Exception =>
  logError("Error starting HiveThriftServer2",e)
          System.exit(-1)
    }
  }
```

7.3 Beeline 操作

Hive 的 Beeline 是第三方 Hive 框架 HiveServer2 提供的一个命令行工具，Beeline 是基于 SQLLine CLI 的 JDBC 客户端。Beeline 工作模式分为本地嵌入模式和远程模式：嵌入模式情况下，Beeline 返回一个嵌入式的 Hive 客户端（类似于 Hive CLI）；远程模式则通过 Thrift 协议与单独的 HiveServer2 进程进行连接通信。

Spark 的 Beeline 是 Spark\bin 目录下的一个应用程序，通过在 spark-class 中执行 org.apache.hive.beeline.Beeline 程序，启动 Beeline 的 JDBC 客户端。当 Thrift Sever 已经启动，我们就可以使用 Beeline 来测试 Spark SQL Thrift JDBC/ODBC server 了。通过 Beeline 命令终端，可以像在 Hive 命令终端一样，执行各种 Hive 语法的 SQL 语句。

7.3.1 Beeline 连接方式

根据 Thrift Server 启动时的传输通道模式，Beeline 的连接方式也不同。分别对应为 HTTP 和 Binary 两种不同的模式。

1. 连接到 Binary 模式的 Thrift Server

下面是 Binary 模式下具体的连接示例。

1）在 Spark 的 bin 目录下执行 beeline 命令：

```
root@ Master:/usr/local/spark/spark-1.6.3-bin-hadoop2.6# ./bin/beeline
Beeline version 1.6.0 by Apache Hive
```

2）在 Beeline 命令终端中输入连接信息：

```
beeline > ! connect jdbc:hive2://localhost:10000
Connecting tojdbc:hive2://localhost:10000
```

3）在 Beeline 命令终端中进行 Hive 的身份认证。

在连接 Thrift Sever 时，使用了 hive metastore，所以还需要进行 Hive 身份认证。我们输入 Hive 中配置的认证用户名：root，密码直接按【Enter】键即可。下面是具体的运行信息：

```
Enter username forjdbc:hive2://localhost:10000:root
Enter password forjdbc:hive2://localhost:10000:
16/04/30 19:45:28 INFOjdbc.Utils:Supplied authorities:localhost:10000
```

```
16/04/30 19:45:28 INFOjdbc.Utils:Resolved authority:localhost:10000
16/04/30 19:45:29 INFOjdbc.HiveConnection:Will try to open client transport with JDBC Uri:jdbc:hive2://localhost:10000
Connected to:Spark SQL(version 1.6.0)
Driver:Spark Project Core(version 1.6.0)
Transaction isolation:TRANSACTION_REPEATABLE_READ
```

当 Hive 身份认证通过之后，提示我们已经正确连接到 Spark SQL。

下面对 Hive 的授权做简单介绍，HiveServer2 支持匿名（非授权）以及 SASL、KERBEROS、LDAP、插件式客户授权和插件式授权模式，如表 7-3 所示。

表 7-3 HiveServer2 授权配置参数说明

函　数	说　明
hive.server2.authentication	授权模式，默认 NONE，可选的包括 NOSASL、KERBEROS、LDAP、PAM 和 CUSTOM 等授权模式 NONE：无身份验证检查 LDAP：基于 LDAP/AD 的身份认证 KERBEROS：Kerberos/GSSAPI 认证 CUSTOM：自定义身份验证提供程序（使用 hive.server2.custom.authentication.class） PAM：可插入的验证模块 NOSASL：Raw transport
hive.server2.authentication.kerberos.principal	被授权的用户． 例如：hive/HiveServer2Host@YOUR-REALM.COM
hive.server2.authentication.kerberos.keytab	用户证书文件为 keytab，例如：/etc/hive/conf/hive.keytab
hive.server2.authentication.ldap.url	LDAP（Lightweight Directory Access Protocol）服务器地址 URL，例如：ldap://myserver:389
hive.server2.authentication.ldap.baseDN	LDAP 的 baseDN 根域路径， 例如：ou=People,dc=my-domain,dc=com
hive.server2.custom.authentication.class	如果将 hive.server2.authentication 设置成 CUSTOM，则需设置 hive.server2.custom.authentication.class 指定权限认证的类，这个类需要实现 　　org.apache.hive.service.auth.PasswdAuthenticationProvider 接口，HiveServer2 调用其 Authenticate（user, passed）认证请求的方法，通过 Hadoop 的 org.apache.hadoop.conf.Configurable 类获取 Hive 的配置对象

可以在 hive 的 conf 目录，对 hive-site.xml 进行配置：

```
<property>
<name>hive.server2.authentication</name>
<value>NONE</value>  // 配置为 NONE,这里不进行身份验证检查
<description>
    Expects one of [nosasl, none, ldap, kerberos, pam, custom].
    Client authentication types.
        NONE: no authentication check//无身份验证检查
```

第7章 Thrift Server

```
        LDAP：LDAP/AD based authentication//基于 LDAP/AD 身份认证
        KERBEROS：Kerberos/GSSAPI authentication //Kerberos/GSSAPI 身份认证
        CUSTOM：Custom authentication provider //自定义认证
                  (Use with property hive.server2.custom.authentication.class)
        PAM：Pluggable authentication module //可插入验证模块
        NOSASL：Raw transport//非简单安全验证
    </description>
</property>
```

如果使用上面的配置，重启 Hive Metastore 服务之后，那么 Beeline 连接到 Thrift Sever 就不用输入用户名和密码，直接按 Enter 键就可以了。但是根据集群实际情况，可能会遇到下面这种错误：

```
org.apache.Hadoop.security.AccessControlException：Permission denied：
Permission denied……
```

只需要在 Hadoop 的配置目录下，增加下面配置即可：

```
<property>
    <name>dfs.permissions</name>
    <value>false</value>
</property>
```

2. 连接到 HTTP 模式的 Thrift Server

Beeline 使用 HTTP 模式连接到 JDBC/ODBC 服务时，连接 URL 的格式如下：

```
beeline>! connect jdbc:hive2://<host>:<port>/<database>? hive.server2.transport.mode=http;hive.server2.thrift.http.path=<http_endpoint>
```

Beeline 连接 URL 的格式参数，请参阅前面的"表 7-1 使用 HTTP 模式的参数表"。

下面是 HTTP 模式下具体的连接示例。

1）在 Spark 的 bin 目录下执行 Beeline 命令：

```
root@ Master:/usr/local/spark/spark-1.6.3-bin-hadoop2.6# ./bin/beeline
Beeline version 1.6.0 by Apache Hive
```

2）在 beeline 命令终端中输入连接信息：

```
beeline>! connect jdbc:hive2://localhost:10001/hive;transportMode=http;httpPath=cliservice
Connecting tojdbc:hive2://localhost:10001/hive;transportMode=http;httpPath=cliservice
```

3）在 Beeline 命令终端中进行 Hive 的身份认证：

```
Enter username forjdbc:hive2://localhost:10001/hive;transportMode=http;httpPath=cliservice:root
```

```
Enter password forjdbc:hive2://localhost:10001/hive;transportMode = http;httpPath = cliservice:
16/05/11 17:47:11 INFOjdbc.Utils:Supplied authorities:localhost:10001
16/05/11 17:47:11 INFOjdbc.Utils:Resolved authority:localhost:10001
Connected to:Spark SQL(version 1.6.0)
Driver:Spark Project Core(version 1.6.0)
Transaction isolation:TRANSACTION_REPEATABLE_READ
0:jdbc:hive2://localhost:10001/hive >
```

同样在进行身份认证时，输入用户名：root，输入密码时，直接按 Enter 键即可。之后会提示正确连接到 Spark SQL。

7.3.2 在 Beeline 中进行 SQL 查询操作

Beeline 与 Thrift Server 连接成功之后，就可以在 Beeline 命令终端中执行 Hive 的语法了，最终的 SQL 查询计算是由 Spark SQL 来完成的，SQL 的查询计算效率比直接在 Hive 中执行提高了几个数量级。

下面演示一下在 Beeline 中进行最基本的 SQL 查询操作。

1）查询列出数据库：

```
0:jdbc:hive2://localhost:10000 > show databases;
+-----------+--+
|  result   |
+-----------+--+
|  default  |
|  hive     |
+-----------+--+
2 rows selected(1.125 seconds)
```

结果为查询 Hive 中的数据库，包括 default 数据库、Hive 数据库。

2）使用某个数据库：

```
0:jdbc:hive2://localhost:10000 > use hive;
+-----------+--+
|  result   |
+-----------+--+
+-----------+--+
No rows selected(0.789 seconds)
```

输入 use hive 命令后，使用相应的数据库。

3）列出数据库中的表：

```
0:jdbc:hive2://localhost:10000 > show tables;
+----------------+----------------+--+
```

```
|   tableName   |  isTemporary  |
+---------------+---------------+--+
|   a1          |  false        |
|   a11         |  false        |
|   people      |  false        |
|   scores      |  false        |
|   scoresresult|  false        |
|   userlogs    |  false        |
+---------------+---------------+--+
6 rows selected(4.893 seconds)
```

结果为查询hive数据库中的所有表。

4) 显示表的建表信息：

```
0:jdbc:hive2://localhost:10000 > show create table people;
+----------------------------------------------------------------------+--+
|                              result                                  |
+----------------------------------------------------------------------+--+
|   CREATE TABLE 'people'(                                             |
|     'name' string,                                                   |
|     'age' int)                                                       |
|   ROW FORMAT DELIMITED                                               |
|     FIELDS TERMINATED BY ','                                         |
|     LINES TERMINATED BY '\n'                                         |
|   STORED AS INPUTFORMAT                                              |
|     'org.apache.hadoop.mapred.TextInputFormat'                       |
|   OUTPUTFORMAT                                                       |
|     'org.apache.hadoop.hive.ql.io.HiveIgnoreKeyTextOutputFormat'     |
|   LOCATION                                                           |
|     'hdfs://Master:9000/user/hive/warehouse/hive.db/people'          |
|   TBLPROPERTIES(                                                     |
|     'COLUMN_STATS_ACCURATE'='true',                                  |
|     'numFiles'='1',                                                  |
|     'totalSize'='71',                                                |
|     'transient_lastDdlTime'='1462785638')                            |
+----------------------------------------------------------------------+--+
17 rows selected(0.464 seconds)
```

5) 查询表的记录：

```
0:jdbc:hive2://localhost:10000 > select * from people limit 10;
```

```
+----------+-----+
|   name   | age |
+----------+-----+
|   Lily   | 18  |
|  Frank   | 30  |
|   Jack   | 20  |
|  Jason   | 50  |
|   Dave   | 35  |
| Stephen  | 28  |
|  Barton  | 21  |
|   Bill   | 22  |
+----------+-----+
8 rows selected(0.593 seconds)
```

通过上面演示，可以看见在 Beeline 中很方便进行 SQL 查询。同样的，在 Hive 和 Spark SQL 上，Spark SQL 会快很多。由此可见，在 Spark 中，Spark SQL 负责高性能的查询计算引擎，Hive 只是作为数据仓库的数据存储，查询计算任务交给 Spark SQL 来完成。

最后说一下，用户可以使用【Ctrl + C】组合键或者使用" !q" 命令退出 Beeline。

7.3.3 通过 Web 控制台查看用户进行的操作

在浏览器中输入 http://master:4040，可以查看用户进行的 JDBC/ODBC 操作，如图 7-4 所示。

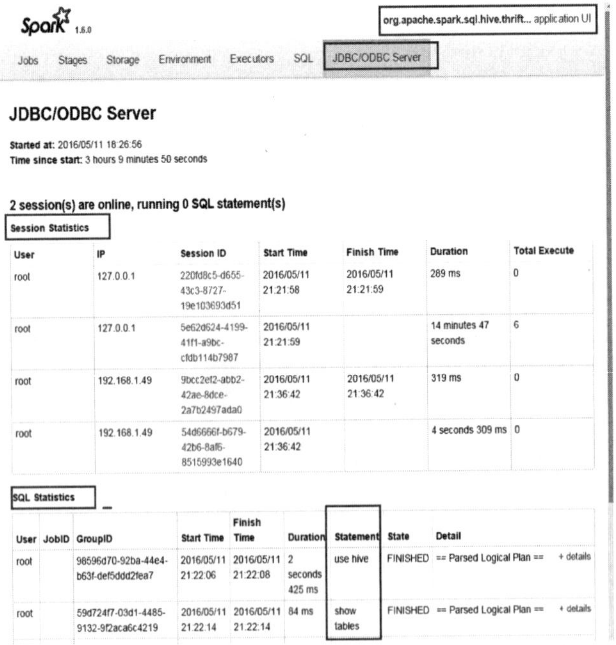

图 7-4 JDBC/ODBC Server

第7章 Thrift Server

上面的监控网页就是HiveThriftServer2应用程序的运行情况,在Session Statistics区域可以查看此时连接的客户端情况,例如,有哪些客户端连接了Thrift Sever,在SQL Statistics区域可以查看具体执行的SQL语句情况,例如use hive, Showtable语句。

其中:
- application UI处显示应用程序名org. apache. spark. sql. hive. thriftserver。
- 在Spark的Tab项(Jobs、Stages、Storage、Environment、Executors、SQL、JDBC/ODBC Server)单击JDBC/ODBC Server栏目,查询JDBC/ODBC Server服务的相关信息。
- 在Session Statistics区域,查询客户端会话的连接。
- 在SQL Statistics区域,查询执行的SQL语句。

7.4 Thrift Server 应用示例

本节讲解Spark Thrift Server的应用示例:在Hive数据库人员信息表people中,查询统计年龄大于30岁的员工信息(姓名、年龄)。

访问Hive数据库表数据有多种方式:
- 通过Hive提供的JDBC接口,直接连接Hiveserver访问Hive数据。例如可以使用Java代码来连接Hive,进行SQL语句查询操作。
- 使用Spark Thrift Server方式,通过JAVA JDBC/ODBC访问Spark Thrift Server, Spark Thrift Server访问Spark SQL,然后通过Spark SQL访问Hive数据库中的数据。
- 本章节使用Spark Thrift Server方式访问,相关的环境准备如下:
- Hive数据库环境。
- Spark集群环境。
- 启动hive --service metastore服务。
- 启动Spark Thrift Server。

7.4.1 示例源代码

```
package com. dt. spark. SparkApps. sql;

import java. sql. Connection;
import java. sql. DriverManager;
import java. sql. PreparedStatement;
import java. sql. ResultSet;
import java. sql. SQLException;

/**
 * 实战演示Java通过JDBC访问Thrift Server,再访问Spark SQL,进而访问Hive,这是企业级开
 *   发中最为常见的方式
```

```java
 *
 */
public class SparkSQLJDBC2ThriftServer{
    public static void main(String[] args){
        String sql = "select name,age from people where age >= ?";
        Connection conn = null;
        ResultSet resultSet = null;
        try{
            Class.forName("org.apache.hive.jdbc.HiveDriver");

            conn = DriverManager.getConnection(
"jdbc:hive2://Master:10001/hive;transportMode=http;httpPath=cliservice","root","");

            java.sql.Statement stmt = conn.createStatement();
            stmt.execute("use hive");

            PreparedStatement preparedStatement = conn.prepareStatement(sql);
            preparedStatement.setInt(1,30);
            resultSet = preparedStatement.executeQuery();

            while(resultSet.next()){
                System.out.println("name:" + resultSet.getString(1) + ",age = "
                    + resultSet.getString(2));
            }

        }catch(Exception e){
            e.printStackTrace();
        }finally{
            try{
                resultSet.close();
                conn.close();
            }catch(SQLException e){
                e.printStackTrace();
            }
        }
    }
}
```

7.4.2 关键代码行解析

下面对本示例的关键代码进行解析。

第7章 Thrift Server

1）首先，加载 HiveServer2 的 JDBC 驱动：

  Class. forName("org. apache. hive. jdbc. HiveDriver");

在使用 Class. forName( )加载 hiveserver2 的驱动 org. apache. hive. jdbc. HiveDriver 时，会执行 HiveDriver 中的静态代码段，创建一个 HiveDriver 实例，然后调用 DriverManager. registerDriver( )注册。

2）建立指定驱动的 URL 连接：

  conn = DriverManager. getConnection(
    "jdbc:hive2://Master:10001/hive;transportMode = http;httpPath = cliservice","root","");

- url：驱动连接 URL。
- user：是授权访问的用户名。
- password：授权访问的密码。

3）连接 URL。

① 当 HiveServer2 使用 HTTP 模式运行时，JDBC 连接 URL 格式如下：

  jdbc:hive2://< host >:< port >/< db >;transportMode = http;httpPath = < http_endpoint >

其中，< http_endpoint > 默认值是：cliservice，需要跟 hive – site. xml 中的配置保持一致。< port > 端口的默认值是：10001。

② 当 HiveServer2 使用 Binary 模式运行时，JDBC 连接 URL 格式如下：

  jdbc:hive2://< host >:< port >/< db >

其中，< port > 端口的默认值是：10000。

4）动态选定数据库。

  stmt. execute("use hive");

在代码中，通过 use DatabaseName 语句即可实现动态选定数据库。

5）组织 SQL 语句。

  Stringsql = "select name,age from people where age >= ?";
  PreparedStatementpreparedStatement = conn. prepareStatement(sql);
  preparedStatement. setInt(1,30);

组织 SQL 语句：从 people 表中查询 age 大于等于 30 的记录。这种写法可以防范 SQL 注入，不用对传入的数据做任何过滤，而如果使用普通的 statement 实例就有 SQL 注入风险。而且 PreparedStatement 实例包含已编译的 SQL 语句，其执行速度要快于 Statement 对象。

6）执行 SQL 语句。

  resultSet = preparedStatement. executeQuery( );

7) 遍历查询结果对象。

```
while( resultSet. next( ) ) {
    System. out. println( "name:" + resultSet. getString(1) + " ,age = " + resultSet. getString(2) ) ;
}
```

8) 本示例并不需要 SparkContext?

在本示例中，为什么不需要 SparkContext 呢？我们通常用 Java 来操作 Spark SQL，肯定有类似下面的代码：

```
SparkConf conf = new SparkConf( ). setMaster( "local" ). setAppName( "xxxx" ) ;
JavaSparkContext sc = new JavaSparkContext( conf) ;
SQLContextsqlContext = new SQLContext( sc) ;
```

原因是：我们只是把 SQL 查询语句指令发给 Thrift Server，然后 Thrift Server 把 SQL 查询语句交给了 Spark SQL 处理，所以就跟普通的 Spark SQL 程序不一样。

7.4.3 测试运行

1. 启动集群环境

1) 在 Hadoop 目录下启动：start-dfs.sh 和 start-yarn.sh，执行命令：

```
root@ Master:/usr/local/hadoop/hadoop-2.7.1/sbin# ./start-all.sh
```

2) 在 Spark 目录下执行命令：

```
root@ Master:/usr/local/spark/spark-1.6.3-bin-hadoop2.6/sbin# ./start-all.sh
```

3) 在 Spark 目录下启动 HistoryServer：

```
root@ Master:/usr/local/spark/spark-1.6.3-bin-hadoop2.6/sbin# ./start-history-server.sh
```

4) 启动 Hive Metastore 服务，命令如下：

```
root@ Master:/usr/local/spark/spark-1.6.3-bin-hadoop2.6# hive --service metastore&
```

2. 启动 Thrift Sever

本案例以 HTTP 通道模式连接 Thrift Sever，所以启动 Thrift Sever 时，请以 HTTP 模式启动，命令如下：

```
root@ Master:/usr/local/spark/spark-1.6.3-bin-hadoop2.6/sbin# ./start-thriftserver.sh --master spark://master:7077 --hiveconf hive.server2.transport.mode=http --hiveconf hive.server2.thrift.http.path=cliservice
starting org. apache. spark. sql. hive. thriftserver. HiveThriftServer2 ,logging to
/usr/local/spark/spark-1.6.3-bin-hadoop2.6/logs/spark-root-org.apache.spark.sql.hive.thriftserver.HiveThriftServer2-1-Master.out
```

3. 在 Hive 中准备数据

为了运行本案例,我们必须在 Hive 数据库中创建一个名为 people 的表,该表有 name、age 两个字段,存储了下面几条测试记录:

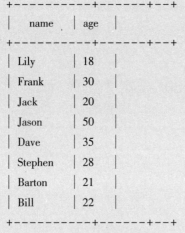

```
0:jdbc:hive2://localhost:10000> select * from people limit 10;
+-----------+------+
|   name    | age  |
+-----------+------+
|   Lily    |  18  |
|   Frank   |  30  |
|   Jack    |  20  |
|   Jason   |  50  |
|   Dave    |  35  |
|  Stephen  |  28  |
|  Barton   |  21  |
|   Bill    |  22  |
+-----------+------+
8 rows selected(0.593 seconds)
```

我们可以通过下面的方法来创建测试数据。

1) 首先创建一个名为 people.txt 的文本文件。

执行命令: vi /mnt/test/people.txt,文件内容如下:

```
Lily,18
Frank,30
Jack,20
Jason,50
Dave,35
Stephen,28
Barton,21
Bill,22
```

2) 连接 Hive 数据库:

```
hive> use hive;
OK
Time taken: 0.041 seconds
```

3) 创建 people 表:

```
hive> CREATE TABLE 'people'('name'string, 'age'int)ROW FORMAT DELIMITED FIELDS TER-
MINATED BY ',' LINES TERMINATED BY '\n';
OK
Time taken: 0.257 seconds
```

4）此时可以查看一下 people 表的描述：

```
hive > desc people;
OK
name                    string
age                     int
Time taken: 0.727 seconds, Fetched: 2 row(s)
```

5）往 pepole 表中装载数据：

```
hive > load data localinpath '/mnt/test/people.txt' into table people;
Loading data to table hive.people
Table hive.people stats: [numFiles = 1, totalSize = 71]
OK
Time taken: 0.256 seconds
```

或者使用下面更简单的方式插入数据：

```
INSERT OVERWRITE TABLE people
select * from(
    select 'Lily' as name,18 as age union all
    select 'Frank',30    union all
    select 'Jack',20     union all
    select 'Jason',50    union all
    select 'Dave',35     union all
    select 'Stephen',28 union all
    select 'Barton',21   union all
    select 'Bill',22
)g;
```

这种方法不需要创建测试数据文件。

6）查询 pepole 表，结果如下：

```
hive > select * from people;
OK
Lily    18
Frank   30
Jack    20
Jason   50
Dave    35
Stephen 28
Barton  21
Bill    22
Time taken: 0.056 seconds, Fetched: 8 row(s)
```

7.4.4 运行结果解析

把本示例的源码文件在 IDE 中打包生成 JAR 包,然后提交到集群上的 Driver 机器上执行,运行命令如下:

```
root@Master:/usr/local/spark/spark-1.6.3-bin-hadoop2.6/sbin#
    /usr/local/spark/spark-1.6.3-bin-hadoop2.6/bin/spark-submit --class
        com.dt.spark.SparkApps.sql.SparkSQLJDBC2ThriftServer --master spark://Master:7077 /
    mnt/job/SparkApps20160504.jar
```

执行结果如下:

```
16/05/09 20:57:33 INFOjdbc.Utils: Supplied authorities: Master:10001
16/05/09 20:57:33 INFOjdbc.Utils: Resolved authority: Master:10001
name:Frank,age=30
name:Jason,age=50
name:Dave,age=35
```

因为在 people 表中,我们只生成了 8 条记录,分别为:

```
Lily     18
Frank    30
Jack     20
Jason    50
Dave     35
Stephen  28
Barton   21
Bill     22
```

所以当我们过滤 age 大于等于 30 的记录时,只有:Frank、Jason、Dave 三条记录满足条件。执行结果符合我们的预期。

7.4.5 Spark Web 控制台查看运行日志

程序运行完毕,在浏览器打开:http://Master:8080,就可以通过 Spark Web 控制台查看该案例的详细运行日志情况,如图 7-6 所示。

在上图的 Running Applications 区域,单击方框中的:org.apache.spark.sql.hive.thriftserver.HiveThriftServer2,即可查看详细的 Jobs 信息,如图 7-7 所示。

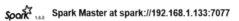

图 7-6　案例运行日志

图 7-7　案例完成的 Jobs

7.5　本章小结

　　本章首先阐述了 Thrift 的发展和基本概念，然后讲解在 Spark SQL 中 Thrift Server 的启动方式，以及 HiveThriftServer2 类的解析，并演示了如何通过 Beeline 来操作 Thrift Sever，最后通过一个 Java 案例来演示 JDBC 下操作 Thrift Sever。通过本章内容，读者可以基本掌握 Thrift Server 的应用。

第8章 Spark SQL综合应用案例

在实际生产环境中，一个电子商务网站每天将会产生海量的日志数据。本章从实战角度详细阐述两个 Spark SQL 的综合应用案例，在电商网站日志多维度数据分析案例中实战演示 UV（Unique Visitor，独立访客数）数据、PV 数据（Page View，页面浏览量）、用户跳出率、新用户注册比例、热门板块排名等分析统计操作；在电商网站搜索排名统计案例中实战演示搜索平台上用户每天搜索排名数据的前 3 名的产品的操作。

Section 8.1 综合案例实战——电商网站日志多维度数据分析

在实际生产环境中，一个电子商务网站在日常运行过程中每天都要产生海量的日志信息，这些日志信息一般主要来源于论坛的日志，网站操作的日志，移动 APP 的日志（iPhone 或 Andriod 手机）等等。

衡量网站日志信息的几个重要参数如下：

1）PV：是 Page View 的简写，即页面浏览量或点击量，用户每 1 次对网站中的每个网页访问均被记录 1 次，也就是 1 个 PV。用户对同一页面的多次访问，访问量累计，也就是统计为多个 PV。

2）UV：是 Unique Visitor 的简写，即独立访客数，是指通过网络访问、浏览某个网页的自然人。比如，在一台电脑上，A 打开了某网站，注册了一个会员，B 通过该电脑注册了另一个会员。由于 A、B 两个人使用的是同一台计算机，那么他们的 IP 地址是一样的，网站日志记录器记录到同一个 IP 地址的登录信息。但是，具有统计分析功能的网站统计系统，可以根据其他条件判断出实际使用的用户数量，返回给网站建设者真实、可信和准确的信息，也就是两个UV。比如，通过注册的用户，甚至可以区分出网吧、机房等共享一个 IP 地址的不同计算机等。

IP 地址是一个反映网络虚拟地址对象的概念，UV 是一个反映实际使用者的概念，每个 UV 相对于每个 IP，更加准确地对应一个实际的浏览者。使用 UV 作为统计量，可以更加准确的了解单位时间内实际上有多少个访问者来到了相应的页面。

对于网站的 PV 和 UV 的统计，可使用第三方统计工具进行统计，将第三方统计工具的代码嵌入至网站中统计 PV 和 UV 的页面，然后登录统计工具后台查询网站的 PV 和 UV 量。

3）BR：是 Bounce Rate 的简写，即用户跳出率，是评价一个网站性能的重要指标，跳出率高，说明网站用户体验做得不好，用户进去就跳出去了，反之如果跳出率较低，说明网站用户体验做得不错，用户能够找到自己需要的内容，而且以后他可能还会再来光顾你的网站，提高了用户黏性，慢慢地可以积累大量的网站用户。

4）板块热度排名：是指在一定时间内用户对网站各板块的浏览量，该参数对于网站总

229

体板块的布局是至关重要的。

5）用户的注册率：是指在一定时间内匿名浏览该网站的用户实际注册为正式用户的比率，一个论坛最好使用活跃用户转化率，一个网上商城最好使用付费用户转化率，一个交友网站可以用注册用户转化率。

8.1.1 数据准备

通常情况下，对日志的处理采用定时任务，在 Linux 环境下，将对日志处理的指令采用 shell 进行封装，通过调度器进行定时调度，常用的调度器有 crontab、ariflow、oozie 等，处理的结果最终放到数据库中。注意：调度器要安装在一台安装了 Spark 客户端的机器上。

8.1.2 数据说明

以下的日志数据来源是用程序模仿网站日志生成的。主要的字段如下：
- 日期，riqi，格式为 yyyy – mm – dd，是指日志记录生成的日期。
- 时间戳，shijian，是指日期记录生成的时间。
- 用户 id，userID，浏览网页的用户，其中 Null 为匿名用户。
- 页面 id，pageID，网站页面的 ID。
- 板块名称，channel，网站板块的名称。
- 操作，action，表明用户是在浏览网页，还是在注册用户。

生成的文件名为 userLogs.txt，纯文本文件，各字段之间用 TAB 分割，每行一条记录。

8.1.3 数据创建

为了读者测试方便，本节将分别采用 Java 和 Scala 语言编写程序实现日志数据生成。

（1）Java 语言生成日志数据

```
import java.io.FileNotFoundException;
import java.io.FileOutputStream;
import java.io.OutputStreamWriter;
import java.io.PrintWriter;
import java.text.SimpleDateFormat;
import java.util.Calendar;
import java.util.Date;
import java.util.Random;
public class SparkSQLDataManally{
    //定义网站具体的板块名称
    static String[] channelNames = new String[]{
        "Spark","Scala","Kafka","Flink","Hadoop","Storm","Hive","Impala",
        "HBase","ML"};
```

第8章 Spark SQL综合应用案例

```
            //定义用户在网页的操作
static String[] actionNames = new String[]{"View","Register"};
static String riqiFormated;
public static void main(String [] args){
        /**
         *通过传递进来的第一个参数生成指定大小规模的数据,默认为5000行;
         *第二个参数指定生成的日志数据文件存放的位置
         */
long numberItems = 5000;
        String path = ".";
if (args.length > 0){
numberItems = Integer.valueOf(args[0]);
path = args[1];
        }
        System.out.println("The total user log number is :" + numberItems + "。");
        //日志时间的生成
riqiFormated = riqi();
userlogs(numberItems,path);
    }
private static void userlogs(long numberItems,String path) {
        Random random = new Random();
String BufferuserLogBuffer = new StringBuffer("");
int[] unregisteredUsers = new int[]{1,2,3,4,5,6,7,8}
for(i <- 1 tonumberItems.toInt)
    {
var   timestamp = new Date().getTime()
var userID = 0L
var pageID = 0L
        //随机生成的用户ID,其中userID为null的是新用户
if unregisteredUsers[random.nextInt(8)] == 1) {
userID = 0L
        } else {
userID = (long)random.nextInt((int) numberItems)
        }
            //随机生成的页面ID
pageID = random.nextInt((int) numberItems);
            //随机生成Channel
            String channel = channelNames[random.nextInt(10)];
            //随机生成action行为
            String action = actionNames[random.nextInt(2)];
userLogBuffer.append(riqiFormated).append("\t")
            .append(timestamp).append("\t")
```

```java
                    .append(userID).append("\t")
                    .append(pageID).append("\t")
                    .append(channel).append("\t")
                    .append(action).append("\n");
        }
        PrintWriter printWriter = null;
        try {
            printWriter = new PrintWriter(new OutputStreamWriter(
                    new FileOutputStream(path + "userLog.log")));
            printWriter.write(userLogBuffer.toString());
        } catch (Exception e) {
            e.printStackTrace();
        } finally {
            printWriter.close();
        }
    }
}
private static String riqi() {
    SimpleDateFormat date = new SimpleDateFormat("yyyy-MM-dd");
    Calendar cal = Calendar.getInstance();
    cal.setTime(new Date());
    cal.add(Calendar.DATE, -1);
    Date yesterday = cal.getTime();
    return date.format(riqi);
}
}
```

（2）Scala 语言生成日志数据

```scala
import java.io.{FileOutputStream, OutputStreamWriter, PrintWriter}
import java.util.Date
import org.joda.time.DateTime
import scala.util.Random

/**
  * Created by qian on 2016/4/16.
  * 网站日志自动生成器代码
  * 日志数据格式：
  * riqi:日期,格式为 yyyy-MM-dd,日志的生成时间
  * timestamp:时间戳
  * userID:用户 ID
  * pageID:页面 ID
  * chanel:板块名称
  * action:浏览(View)或注册(Register)
```

第8章　Spark SQL综合应用案例

```
    */
object SparkSQLDataManually {
//网站的具体频道
    val channelNames = Array("spark","scala","kafka","Flink","hadoop","Storm","Hive","Impala","HBase","ML")
//用户的操作
val actionNames = Array("View","Register")
    var riqiFormated = ""

def main(args:Array[String]) {
//生成日志的数量,命令行的第一个参数确定生成日志的行数,默认为500行
var numberItems:Int = 500
    var path = ""
if(args.length > 0)
    {
numberItems = args(0).toInt
path = args(1)
    }
println("User log number is :" + numberItems)
    //日志时间生成
riqiFormated = DateTime.now.minusDays(1).toString("yyyy-MM-dd")
userlogs(numberItems,path)
    }
//生成网站日志的函数,传入参数有:生成日志的行数和日志的存放路径
def userlogs(numberItems:Long,path:String):Unit = {
var userLogBuffer = new StringBuffer("")
val random = new Random()
    //随机生成的页面ID
pageID = random.nextInt(numberItems.toInt)
        //随机生成的ChannelID
val channel = channelNames(random.nextInt(10))
        //随机生成Action行为
val action = actionNames(random.nextInt(2))
userLogBuffer.append(yesterdayFormated).append("\t")
        .append(timestamp).append("\t")
.append(userID).append("\t")
.append(pageID).append("\t")
        .append(channel).append("\t")
.append(action).append("\n")
    }
val printWriter = new PrintWriter(new OutputStreamWriter(new FileOutputStream(path + "userLog.log")))
try{
```

```
            printWriter.write(userLogBuffer.toString());
        }
    catch {
        case e:Exception = > println(e.toString);
        }
    finally {
        printWriter.close();
        }
    }
}
```

在 IntelliJ IDEA 集成开发环境中，输入快捷键 ALT + Shift + F10，弹出 Run 运行页面，在 program arguments 项输入产生的日志条数和日志文件的存放位置，点击 Run 按钮运行程序。

程序运行后，生成数据的格式如下：

日期时间戳	用户ID	页面ID	板块名称	浏览或注册
2016-04-24	1461578786916	38392	54157	Impala	View
2016-04-24	1461578786916	651242	260613	Spark	View
2016-04-24	1461578786916	437591	720600	Kafka	Register
2016-04-24	1461578786916	0	699510	Spark	View
2016-04-24	1461578786916	225827	317728	Storm	Register
2016-04-24	1461578786916	415604	376904	Flink	Register
2016-04-24	1461578786916	939459	389381	Spark	Register
2016-04-24	1461578786916	50944	920496	Spark	Register
2016-04-24	1461578786916	77695	192414	HBase	Register
2016-04-24	1461578786916	0	489529	ML	View

如果生成的数据量偏大，比如超过 100 万行，即在运行的配置参数页面中设置 program arguments 配置参数中输入 1000000 c:\，如下图 8-1 所示。

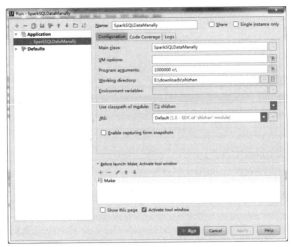

图 8-1　输入运行参数

第8章 Spark SQL综合应用案例

点击 Run 按钮运行程序，将报错：

```
User log number is :10000000
Exception in thread "main" java.lang.OutOfMemoryError:Java heap space
```

出错的原因是 Java 堆空间不足，产生溢出，处理方法是手动调整 JVM 的参数。在运行的配置参数页面中将 VM options 设置参数 -verbose：gc -Xms4G -Xmx4G -Xss2G，如下图 8-2 所示。其含义为：-verbose：gc，在虚拟机发生 GC 垃圾回收时在输出设备显示信息；-Xmx4G，JVM 启动初始化堆大小为 4GB，Xmx4G，JVM 最大的堆大小为 4GB；-Xss2G，每个线程的堆栈大小为 2GB。

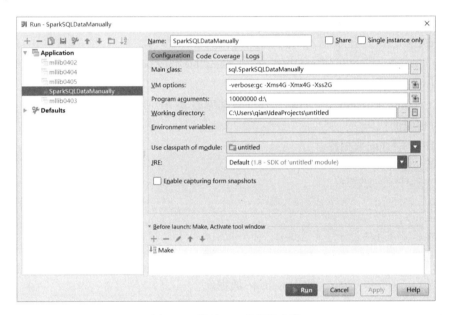

图 8-2 设置 VM 虚拟机参数

点击 Run 按钮运行程序即可生成测试用数据。

8.1.4 数据导入

本节将生成模拟日志数据，记录行数为 100 万行，日志大小约为 500 MB，然后我们需将模拟日志数据导入到 Hive 中，数据导入 Hive 的三种不同方式：

- 在 Hive 中操作加载数据库表 userLogs 的数据。
- 在 Spark Shell 中使用 sqlContext 操作加载数据到 Hive 的表 userLogs 中。
- 使用 spark-sql 应用工具将数据导入 hive 的表 userLogs。

（1）方式一：在 Hive 中操作加载数据库表 userLogs 的数据

第一步：将 userLog.log 文件复制到 hadoop 集群的本地机器的 /home/hadoop/ 目录；

第二步：通过 Hive 将日志数据导入到 Hive 数据仓库中。

1）启动 Hive MetaStore。

```
[hadoop@ master ~ ]$hive --service metastore&
[1] 4186
```

2）启动 hive。

```
[hadoop@ master ~ ]$hive
……
hive >
```

3）在 hive 表 userLogs 中加载测试数据。
- show databases：查询 hive 数据库。
- use website：使用 website 数据库。
- descuserLogs：查看 userLogs 表结构信息。
- load data local inpath '/home/hadoop/userLog.log' into table userLogs：加载本地数据文件到 hive 数据库 userLogs 表中。
- select count(*) from userLogs：查询数据库表 userLogs 的记录行数。

以下为相应的代码。

① 在 Hive 中查询数据库。

```
hive > show databases;
OK
default
website
Time taken:1.249 seconds,Fetched:2 row(s)
```

② 输入 use website 命令，使用 website 数据库。

```
hive > use website;
OK
Time taken:0.029 seconds
```

③ 在 website 数据库中创建表 userLogs：

```
hive > create table userLogs ( riqi string,shijian bigInt,userID bigInt,pageIDbigInt,channel string,actionString) row format delimited fields terminated by '\t' lines terminated by '\n';
OK
Time taken:0.539 seconds
```

④ 查询表 userlogs 的表结构。

```
hive > desc  userLogs;
OK
riqi                 string
shijian              bigint
userid               bigint
```

第8章 Spark SQL综合应用案例

```
pageid                bigint
channel               string
action                string
Time taken:0.088 seconds,Fetched:6 row(s)
```

⑤ 加载本地数据文件到 userlogs 表。

```
hive > load data local inpath '/home/hadoop/userLog.log'into table userLogs;
Loading data to table website.userlogs    //加载数据到 website userlogs 表
Table website.userlogs stats:[numFiles = 1,totalSize = 51582540]
OK
Time taken:0.867 seconds
```

⑥ 查询 userlogs 表的记录数。

```
hive > select count( * ) from userLogs;
……
Total MapReduce CPU Time Spent:0 msec
OK
1000000
Time taken:3.181 seconds,Fetched:1 row(s)
```

(2) 方式二：在 Spark Shell 中使用 sqlContext 操作加载数据到 Hive 的表 userLogs 中。

1) 启动 sparkShell。

```
Last login:Tue Apr 26 05:49:24 2016 from 60.216.201.197
[hadoop@ master ~ ]$spark - shell -- master = spark://master:7077
……
Welcome to
      ____              __
     / __/__  ___ _____/ /__
    _\ \/ _ \/ _ `/ __/  '_/
   /___/ .__/\_,_/_/ /_/\_\   version 1.6.1
      /_/

Using Scala version 2.10.5 (Java HotSpot(TM) 64 - Bit Server VM,Java 1.8.0_73)
……
16/04/26 05:51:56 INFO repl.SparkILoop:Created sql context (with Hive support)..
SQL context available as sqlContext.
```

日志中显示 "Created sql context (with Hive support)"，表明 sqlContext 已经默认支持 Hive。Spark 中的日志级别对应于 log4j 的日志级别，优先级从高到低依次为：OFF、FATAL、ERROR、WARN、INFO、DEBUG、TRACE、ALL。为方便查看数据结果，这里设置日志级别为 "WARN"，简化 Spark 的日志输出：

```
sc.setLogLevel("WARN")
```

2)在 Spark Shell 导入数据。
- sqlContext.sql("show databases").show:查询数据库。
- sqlContext.sql("use website"):使用 website 数据库。
- sqlContext.sql("show tables").show:查询数据库中的表。
- sqlContext.sql("drop table userLogs"):删掉数据库中表 userLogs。
- sqlContext.sql("load data local inpath '/home/hadoop/userLog.log' into table userLogs"):加载本地数据文件到数据库表 userLogs 中。
- sqlContext.sql("select count(*) from userLogs").show:查询数据库表 userLogs 的记录数。

```
scala > sqlContext.sql("show databases").show;
+-------+
| result|
+-------+
|default|
|website|
+-------+
scala > sqlContext.sql("use website");
res2:org.apache.spark.sql.DataFrame = [result:string]
scala > sqlContext.sql("show tables").show;
+---------+-----------+
|tableName|isTemporary|
+---------+-----------+
| userlogs|      false|
+---------+-----------+
scala > sqlContext.sql("drop table userLogs")
scala > sqlContext.sql("create table userLogs (riqi string,shijian bigInt,userID bigInt,pageIDbigInt,channel string,action String) row format delimited fields terminated by '\t' lines terminated by '\n'");
scala > sqlContext.sql("load data local inpath '/home/hadoop/userLog.log' into table userLogs");
scala > sqlContext.sql("select count(*) from  userLogs").show;
+-------+
|    _c0|
+-------+
|1000000|
+-------+
```

(3)方式三:使用 Spark-Sql 应用工具将数据导入 hive 的表 userLogs。通过 Spark SQL,将数据导入到 hive 数据仓库中
- spark-sql:启动 spark-sql 应用,以下操作在 spark-sql 中执行。

第8章 Spark SQL综合应用案例

- show databases：查询数据库。
- use website：使用数据库 website。
- show tables：查询数据库表。
- drop table userLogs：删除数据表 userLogs。
- create table userLogs：创建数据库表 userLogs。
- load data local inpath '/home/hadoop/userLog.log' into table userLogs：加载本地文件到数据库表 userLogs。
- select count(*) from userLogs：查询数据库表 userLogs 的记录数。

```
[hadoop@master ~]$spark-sql --master=spark://master:7077
SET spark.sql.hive.version=1.2.1
SET spark.sql.hive.version=1.2.1
spark-sql> show databases;
default
website
Time taken:4.237 seconds,Fetched 2 row(s)
16/04/26 06:21:08 INFO CliDriver:Time taken:4.237 seconds,Fetched 2 row(s)
spark-sql> use website;
Time taken:0.139 seconds
16/04/26 06:22:24 INFO CliDriver:Time taken:0.139 seconds
spark-sql> show tables;
userlogs          false
Time taken:0.636 seconds,Fetched 1 row(s)
16/04/26 06:23:17 INFO CliDriver:Time taken:0.636 seconds,Fetched 1 row(s)
spark-sql> drop table userLogs;
Time taken:0.771 seconds
16/04/26 06:24:07 INFO CliDriver:Time taken:0.771 seconds
spark-sql> create table userLogs (riqi string, shijian bigInt, userID bigInt, pageIDbigInt, channel string,action String) row format delimited fields terminated by '\t' lines terminated by '\n';
Time taken:0.499 seconds
16/04/26 06:25:40 INFO CliDriver:Time taken:0.499 seconds
spark-sql> load data local inpath '/home/hadoop/userLog.log' into table userLogs;
Time taken:0.963 seconds
16/04/26 06:26:34 INFO CliDriver:Time taken:0.963 seconds
spark-sql> select count(*) from userLogs;
1000000
Time taken:4.479 seconds,Fetched 1 row(s)
16/04/26 06:27:14 INFO CliDriver:Time taken:4.479 seconds,Fetched 1 row(s)
spark-sql>
```

通过执行上述代码，可查询验证表 userlogs 中有 1000000 条记录。

8.1.5 数据测试和处理

在上一节中已经演示了将测试数据采用不同的方式导入到 Hive 数据仓库中，在本节中所有的数据处理操作均在 Spark SQL 环境下实现。

1. 统计 UV 数据

UV 是指一个页面被多少用户访问过，一个用户多次访问同一页面为 1 个 UV。

下面的代码显示了，针对某天（2016 年 4 月 24 日）浏览网页的用户，UV 数最多的 10 个用户的用户 ID 和 UV 数：

```
spark-sql> select pageID,count(distinct userID) UV from userLogs where action ='View' and riqi =
  '2016-04-24' group by pageID  order by UV desc limit 10;
973122    7    //第一列含义是页面 ID;第二列含义是唯一不重复用户 ID 的 UV 访问累计总数
292220    7
212271    6
161650    6
511705    6
303446    6
458256    6
280080    6
492301    6
18050     6
Time taken:6.364 seconds,Fetched 10 row(s)
16/04/27 05:50:55 INFOCliDriver:Time taken:6.364 seconds,Fetched 10 row(s)
spark-sql>
```

2. 统计 PV 数据

用户每 1 次对网站中的某个网页的访问均被记录 1 次，也就是 1 个 PV。用户对同一页面的多次访问，访问量累计，也就是统计为多个 PV。

下面的代码为，在某天（2016 年 4 月 24 日），网页浏览次数（PV）统计量最多的前 10 个排名。

```
select riqi,pageID,pv from (select riqi,pageID,count(*) pv from userLogs where ation ='View' and
        riqi ='2016-04-24' group by riqi,pageID) subquery order by pv desc limit 10;
    2016-04-24    973122    7
    2016-04-24    292220    7
    2016-04-24    842569    6
    2016-04-24    315316    6
    2016-04-24    973344    6
    2016-04-24    492301    6
    2016-04-24    140524    6
    2016-04-24    224728    6
```

第8章 Spark SQL综合应用案例

```
2016 - 04 - 24         1127      6
2016 - 04 - 24         511705    6
Time taken:8.842 seconds,Fetched 10 row(s)
16/04/27 06:39:23 INFO CliDriver:Time taken:8.842 seconds,Fetched 10 row(s)
spark - sql >
```

3. 用户跳出率统计

用户跳出率是分析网站性能的重要指标,是指用户只访问了入口页面(例如网站首页)就离开的访问量与所产生总访问量的百分比。本节用户跳出章计算方法为:只浏览了一次网站的用户与浏览网站的总用户数的比。下面的代码为先计算某天(2016 年 4 月 24 日)浏览网站的总用户数。

```
scala > val allUser = sqlContext.sql("select count(distinctuserID)  from userLogs where Ation ='View'
    and riqi ='2016 - 04 - 24'").collect;
allUser:Array[org.apache.spark.sql.Row] = Array([354685])
```

再计算只浏览了一次网站的用户数。

```
scala > val jumpUser = sqlContext.sql("select count( * ) from (select count( * ) totalNumber from
    userLogs where Ation ='View'and riqi ='2016 - 04 - 24'group by userID having totalNumber =
    1) target").collect;
jumpUser:Array[org.apache.spark.sql.Row] = Array([282778])
```

最后,计算出用户的跳出率。

```
scala > valjumpRate = jumpUser(0).get(0).toString.toDouble/allUser(0).get(0).toString. toDou-
    ble;
jumpRate:Double = 0.7972651789616139
```

通过将 Double 类型转换为 BigDecimal 类型可以计算更高精度的用户跳出率:

```
scala > BigDecimal.valueOf(jumpUser(0).get(0).toString.toDouble)/BigDecimal.valueOf(allUser
    (0).get(0).toString.toDouble);
res8:scala.math.BigDecimal = 0.7972651789616138263529610781397578
```

当然,在实际环境中,肯定要针对往往某个模块或某个网页进行更加精准的用户跳出率的计算,以便网站的设计者对该网站的模块或网页进行优化处理。

4. 统计新用户注册比例

网站的新用户注册比例,也就是该网站新用户的访问数与每天注册用户数的比率。

首先计算每天注册的用户数:

```
scala > val allUser = sqlContext.sql("select count(distinct userID) from  userLogs where action ='Reg-
    ister'and riqi ='2016 - 04 - 24'").collect;
allUser:Array[org.apache.spark.sql.Row] = Array([354560])
```

接着,计算网站新用户的访问数,在本例中也就是用户 ID 为 0 的用户数:

```
scala > val newUser = sqlContext.sql("select count( * ) from  userLogs where userID ='0'and riqi =
       '2016 - 04 - 24'").collect;
newUser:Array[org.apache.spark.sql.Row] = Array([124577])
```

最后,计算网站的新用户注册比例:

```
scala > newUser(0).get(0).toString.toDouble/allUser(0).get(0).toString.toDouble
res12:Double = 0.3513566110108303
```

5. 统计热门板块排名

本节案例使用 Java 实现了热门板块的排名(例如:网站板块可以设计为"大数据""人工智能""云计算"等,热门板块的排名可统计哪些网站板块最热门,点击率最高。)同时也使用 Java 实现了 PV、UV、页面跳出率、新用户注册比例统计的功能。

具体步骤如下:

1) 初始化 JavaSparkContext 及 HiveContext。
2) pvStatistic 函数:计算 PV。
- 进行子查询 subquery,根据日期、页面 ID 分组,从数据库表 userlogs 中查询昨天的浏览记录,包括日期、页面 ID、pv 数等;
- 从子查询 subquery 中查询,根据 pv 数降序排序,最终查询日期、页面 ID、pv 数等内容。

3) hotChannel 函数:计算热门板块。
- 进行子查询 subquery,根据日期、板块分组,从数据库表 userlogs 中查询昨天的浏览记录,包括日期、板块、板块点击数 channelpv 等内容;
- 从子查询 subquery 中查询,根据板块点击数 channelpv 降序排序,最终查询日期、板块、板块点击数 channelpv 等内容。

4) uvStatistic 函数:计算 UV 数。
- 进行子查询 subquery,根据日期、页面 ID、用户 ID 分组,从数据库表 userlogs 中查询昨天的浏览用户,记录包括日期、页面 ID、用户 ID 等内容;
- 进行子查询 result,根据日期、页面 ID 分组,从子查询 subquery 中查询日期、页面 ID、UV 数等;
- 从子查询 result 中查询,根据 UV 数降序排序,最终查询日期、页面 ID、UV 数等内容。

5) jumpOutStatistic 函数:计算页面跳出率。
- 统计昨天浏览网页用户数;
- 进行子查询,根据用户 ID 分组,查询昨天浏览网页且只浏览 1 次的用户数。
- 然后从子查询中统计昨天只浏览 1 次的总用户数;
- 将只浏览 1 次的总用户数除以浏览网页用户数计算出用户跳出率。

6) newUserRegisterPercentStatistic 函数:计算新用户注册的比例。
- 统计昨天浏览网页及用户 ID 为空的用户数;
- 统计昨天注册的用户数;
- 将注册用户数除以浏览用户数计算出新用户注册比例。

第8章　Spark SQL综合应用案例

```java
package sql;

import java.text.SimpleriqiFormat;
import java.util.Calendar;
import java.util.Date;
import org.apache.spark.SparkConf;
import org.apache.spark.api.java.JavaSparkContext;
import org.apache.spark.sql.hive.HiveContext;

/**
 * Table in hive database creation:
 * sqlContext.sql("create table userlogs (riqi string, timestamp bigint, userID bigint, pageID bigint, channel string, action string) ROW FORMAT DELIMITED FIELDS TERMINATED BY '\t' LINES TERMINATED BY '\n'")
 * @author root
 */
class Spark SQLUserlogsOPS {
    public static void main(String[] args) {
        SparkConf conf = new SparkConf().setMaster("spark://123.233.246.100:7077").setAppName("Spark SQLUserlogsOps");
        JavaSparkContext sc = new JavaSparkContext(conf);
        HiveContext hiveContext = new HiveContext(sc.sc());
        String yesterday = "2016-05-03";
        pvStatistic(hiveContext, yesterday);         //PV
        hotChannel(hiveContext, yesterday);  //热门板块
        uvStatistic(hiveContext, yesterday);         //UV
        jumpOutStatistic(hiveContext, yesterday);   //页面跳出率
        newUserRegisterPercentStatistic(hiveContext, yesterday);    //新用户注册的比例
    }
    //计算新用户注册比例
    private static void newUserRegisterPercentStatistic(HiveContexthiveContext, String yesterday) {
        hiveContext.sql("use website");
        String newUserSQL = "SELECT count(1) from userlogs where action ='View' AND riqi ='" + yesterday + "'and userID is NULL";
        String yesterdayRegistered = "SELECT count(1) from userlogs where action ='Register' AND riqi ='" + yesterday + "'";
        Object totalPv = hiveContext.sql(yesterdayRegistered).collect()[0].get(0);
        Object pv2One = hiveContext.sql(newUserSQL).collect()[0].get(0);
        double total = Double.valueOf(totalPv.toString());
        double pv21 = Double.valueOf(pv2One.toString());
        System.out.println("模拟新用户注册比例:" + pv21 / total);
```

```java
        }
//计算页面跳出率
private static void jumpOutStatistic(HiveContext hiveContext,String yesterday){
hiveContext.sql("use website");
        String totalPvSQL = "SELECT count(1) from userlogs where action ='View' AND riqi
='" + yesterday +"'";
            String pv2OneSQL = "SELECT count(1) from (SELECT count(1) pvPerUser FROM
userlogs where action ='View' AND riqi ='" + yesterday +"' GROUP BY userID HAVING pvPerUser = 1) result";
            Object totalPv = hiveContext.sql(totalPvSQL).collect()[0].get(0);
            Object pv2One = hiveContext.sql(pv2OneSQL).collect()[0].get(0);
double total =     Double.valueOf(totalPv.toString());
double pv21 =     Double.valueOf(pv2One.toString());
        System.out.println("跳出率为:" + pv21 / total);
    }
//计算 UV 数。UV 是指某个页面被多少用户访问过
private static void uvStatistic(HiveContext hiveContext,String yesterday){
hiveContext.sql("use website");
        String sqlText = "SELECT riqi,pageID,uv" +
" FROM (SELECT riqi,pageID,count(1) uv " +
" FROM ( SELECT riqi,pageID,userID  FROM userlogs  "
+" WHERE action ='View' AND riqi ='" + yesterday +"' GROUP BY riqi,pageID,userID) subquery GROUP BY riqi,pageID) result "
+" ORDER BY uv DESC ";
hiveContext.sql(sqlText).show();
        }
//计算热门板块排名
private static void hotChannel(HiveContext hiveContext,String yesterday){
hiveContext.sql("use website");
        String sqlText = "SELECT riqi,channel,channelpv" +
" FROM ( SELECT riqi,channel,count(1) channelpv FROM userlogs  "
+" WHERE action ='View' AND riqi ='" + yesterday +"' GROUP BY riqi,channel) subquery"
+" ORDER BY channelpv DESC ";
hiveContext.sql(sqlText).show();
    }
//计算 PV 页面浏览量或点击量
private static void pvStatistic(HiveContext hiveContext,String yesterday){
hiveContext.sql("use website");
        String sqlText = "SELECT riqi,pageID,pv" +
" FROM ( SELECT riqi,pageID,count(1) pv FROM userlogs   "
+" WHERE action ='View' AND riqi ='" + yesterday +"' GROUP BY riqi,pageID) subquery"
+" ORDER BY pv DESC ";
```

第8章 Spark SQL综合应用案例

```
        hiveContext.sql(sqlText).show();
        //把执行结果放在数据库或者 Hive 表中
        }
        }
```

通过上述代码，读者可通过执行上述代码，查询验证各项排名结果。

Section 8.2 综合案例实战——电商网站搜索排名统计

8.2.1 案例概述

本节介绍一个电商网站搜索排名统计的综合案例：以京东为例，用户登录京东网站，在搜索栏中输入搜索词，然后点击搜索按钮，就能在京东网站搜索用户需要的商品。在搜索栏中输入搜索词时，当用户输入第一个词的时候，京东就能根据用户的点击商品搜索排名，自动在搜索栏下拉列表中显示搜索热词，帮助用户快捷的点击需搜索的商品。在本案例中，将实现和京东搜索类似的功能，根据用户搜索词的日志记录，将用户每天搜索排名前3名的商品列出来，系统后台可以将搜索排名记录持久化到数据库中，提供给 Web 系统或其他应用使用。这里将搜索排名前3名的记录保存到磁盘文件系统中，以 JSON 格式保存。

网站搜索综合案例代码分2个模块：

（1）数据生成模块：模拟数据的生成可以使用爬虫代码程序，从网络上爬取相应的用户搜索数据，进行 ETL 数据清理。为简化数据爬取和清洗过程，我们采用模拟生成数据的方式，根据综合案例的数据需求，人工生成模拟数据文件，实现同样类似的功能。

（2）网站搜索排名：找出用户每天搜索排名前3名的产品。

8.2.2 数据准备

在本案例中，根据项目的需求，需要获取用户搜索的日期、用户名称、用户搜索的商品、用户所在的城市、用户通过什么渠道登录网站（例如：安卓手机、苹果手机或平板电脑）等数据信息。模拟数据的字段包括：日期（date），用户（userID），商品（ItemID），城市（CityID），终端（Device）；模拟数据记录数可以任意指定（如万、千万、亿），在代码中赋值给 numberItems 就可以生成 numberItems 条记录，这样在 Spark SQL 中测试数据就非常方便，也符合实际生产应用场景的需求，因此本案例具备生产系统应用的价值。

1. 数据说明

通过程序 Spark SQLUserlogsHottestDataManually 模拟生成数据（程序获取地址：Http://

blog. csdn. net/dnan_2hihna/article/details/78821810）。

日志的字段为：

日期，date，格式为 yyyy – mm – dd，是指日志记录生成的日期

用户 id，userID，在网站进行搜索的用户 ID

商品 id，ItemID，用户在网站搜索的商品名称 ID

城市 id，CityID，用户在哪个城市登录网站

终端，Device，用户通过什么终端渠道登录网站

生成的文件路径为 G：\Spark SQLData\Spark SQLUserlogsHot. log，各字段之间 TAB 分割，每行一条记录。

2. 数据创建

在 Spark SQLUserlogsHottestDataManually. java 应用程序中，通过 numberItems 指定 10000，先生成 10000 条数据，调用 ganerateUserLogs 函数，在 ganerateUserLogs 函数分别调用生成当前日期、随机用户 ID、随机商品 ID、随机城市名称、随机设备类型的各个函数，然后将生成的时间、用户 id、商品、地点、设备信息记录拼接成字符串，将字符串记录保存到磁盘文件系统。

（1）ganerateUserLogs 函数调用

```java
public static void main(String[] args) {
    long numberItems = 10000
        ganerateUserLogs(numberItems,"G:\\Spark SQLData\\");
}
    private static void ganerateUserLogs(long numberItems,String path)

        StringBuffer userLogBuffer = new StringBuffer()
        String filename = "Spark SQLUserlogsHot. log";
        // 元数据:Date、UserID、Item、City、Device;
        for (int i = 0; i < numberItems; i ++)  {
            String date = getCountDate(null,"yyyy – MM – dd", -1);//获取日期
            String userID = ganerateUserID();//随机生成用户 ID
            String ItemID = ganerateItemID();//随机生成商品 ID
            String CityID = ganerateCityIDs();//随机生成城市名称
            String Device = ganerateDevice();//随机生成设备类型
        userLogBuffer. append(date + "\t" + userID + "\t" + ItemID + "\t" + CityID + "\t" + Device + "\n");
            WriteLog(path,filename,userLogBuffer + "");//保存到磁盘文件
        }
    }
```

（2）获取当前日期 getCountDate

模拟生成用户点击搜索的当前日期。

```java
public static String getCountDate(String date,String patton,int step)  {
```

第8章 Spark SQL综合应用案例

```
        SimpleDateFormatsdf = new SimpleDateFormat(patton);//定义日期的格式
        Calendar cal = Calendar.getInstance();//获取Calendar的一个实例
        if(date! = null){
            try{
                cal.setTime(sdf.parse(date));
            }catch(ParseException e){
                e.printStackTrace();
            }
        }
        cal.add(Calendar.DAY_OF_MONTH,step);
        return sdf.format(cal.getTime());//返回当前日期
    }
```

（3）随机生成用户 UserID

定义一个用户 ID 数组，使用随机数从数组中随机获取用户 ID。

```
    private static String ganerateUserID(){
        Random random = new Random();//定义一个随机数
        String[] userID = {// 模拟定义用户 ID
"98415b9c-f3d4-45c3-bc7f-dce3126c6c0b","7371b4bd-8535-461f-a5e2-c4814b2151e1",
"49852bfa-a662-4060-bf68-0ddde5feea1","8768f089-f736-4346-a83d-e23fe05b0ecd",
"a76ff021-049c-4a1a-8372-02f9c51261d5","8d5dc011-cbe2-4332-99cd-a1848ddfd65d",
"a2bccbdf-f0e9-489c-8513-011644cb5cf7","89c79413-a7d1-462c-ab07-01f0835696f7",
"8d525daa-3697-455e-8f02-ab086cda7851","c6f57c89-9871-4a92-9cbe-a2d76cd79cd0",
"19951134-97e1-4f62-8d5c-134077d1f955","3202a063-4ebf-4f3f-a4b7-5e542307d726",
"40a0d872-45cc-46bc-b257-64ad898df281","b891a528-4b5e-4ba7-949c-2a32cb5a75ec",
"0d46d52b-75a2-4df2-b363-43874c9503a2","c1e4b8cf-0116-46bf-8dc9-55eb074ad315",
"6fd24ac6-1bb0-4ea6-a084-52cc22e9be42","5f8780af-93e8-4907-9794-f8c960e87d34",
"692b1947-8b2e-45e4-8051-0319b7f0e438","dde46f46-ff48-4763-9c50-377834ce7137"};
        return userID[random.nextInt(20)];// 随机获取用户 ID
    }
```

（4）随机生成商品 ItemID

定义一个商品 ID 数组，使用随机数从数组中随机获取商品名称 ID。

```
    private static String ganerateItemID(){
        Random random = new Random();
        String[] ItemIDs = {"小米","休闲鞋","洗衣机","显示器","显卡","洗衣液","行车记录仪"};
        return ItemIDs[random.nextInt(7)];
    }
```

（5）随机获取城市名称

定义一个城市名称数组，使用随机数从数组中随机获取城市名称

```
private static StringganerateCityIDs() {
    Random random = new Random();
    String[] CityNames = {"上海","北京","深圳","广州","纽约","伦敦","东京","首尔","莫斯科","巴黎"};
    return CityNames[random.nextInt(10)];
}
```

（6）获取设备类型

定义设备类型数组，使用随机数从数组中随机获取设备类型，如安卓手机、苹果手机、平板电脑

```
private static String ganerateDevice() {
    Random random = new Random();
    String[] Devices = {"android","iphone","ipad"};
    return Devices[random.nextInt(3)];
}
```

（7）写入磁盘文件

将生成的用户搜索元数据记录信息：日期（date），用户（userID），商品（ItemID），城市（CityID），终端（Device）保存到磁盘文件中。

```
public static void WriteLog(String path,String filename,String strUserLog) {
    FileWriter fw = null;
    PrintWriter out = null;
    try {
        File writeFile = new File(path + filename);
        if (!writeFile.exists())
            writeFile.createNewFile();
        else {
            writeFile.delete();
        }
        fw = new FileWriter(writeFile,true);
        out = new PrintWriter(fw);
        out.print(strUserLog);
    } catch (Exception e) {
        e.printStackTrace();
        try {
            if (out! = null)
                out.close();
```

第8章 Spark SQL综合应用案例

```
                        if (fw! = null)
                            fw.close();
                    } catch (IOException ex) {
                            ex.printStackTrace();
                    }
                }
            } finally {
                try {
                    if (out! = null)
                        out.close();
                    if (fw! = null)
                        fw.close();
                } catch (IOException e) {
                        e.printStackTrace();
                }
            }
        }
    }
```

3. 生成模拟数据

运行 Spark SQLUserlogsHottestDataManually.java 代码程序,生成的模拟数据结果保存到 windows 系统中的 "G:\Spark SQLData\Spark SQLUserlogsHot.log" 文件。其中的每行数据分别为日期(date)、用户(userID)、商品(ItemID)、城市(CityID)、终端(Device),从日志文件中选取 10 条记录,运行结果如下。

```
2016 - 08 - 15    89c79413 - a7d1 - 462c - ab07 - 01f0835696f7    小米 莫斯科   iphone
2016 - 08 - 15    692b1947 - 8b2e - 45e4 - 8051 - 0319b7f0e438    显卡 深圳 iphone
2016 - 08 - 15    c6f57c89 - 9871 - 4a92 - 9cbe - a2d76cd79cd0    显示器 上海 ipad
2016 - 08 - 15    49852bfa - a662 - 4060 - bf68 - 0dddde5feea1    小米 巴黎 android
2016 - 08 - 15    a76ff021 - 049c - 4a1a - 8372 - 02f9c51261d5    小米 首尔 android
2016 - 08 - 15    0d46d52b - 75a2 - 4df2 - b363 - 43874c9503a2    行车记录仪 深圳 iphone
2016 - 08 - 15    6fd24ac6 - 1bb0 - 4ea6 - a084 - 52cc22e9be42    显示器 巴黎 iphone
2016 - 08 - 15    c6f57c89 - 9871 - 4a92 - 9cbe - a2d76cd79cd0    小米 深圳 android
2016 - 08 - 15    5f8780af - 93e8 - 4907 - 9794 - f8c960e87d34    洗衣液 深圳 android
2016 - 08 - 15    c6f57c89 - 9871 - 4a92 - 9cbe - a2d76cd79cd0    小米 莫斯科 android
```

至此,本案例的模拟数据已生成,接下来我们实现用户每天搜索前 3 名的商品排名统计。

8.2.3 实现用户每天搜索前 3 名的商品排名统计

Spark SQL 网站搜索综合实战案例实现用户每天搜索前 3 名的商品排名统计,具体实现步骤如下:

1) 从 Spark 读入用户搜索日志记录,根据项目需求进行 ETL 数据清洗。在实际生产应用场景中，过滤条件可能非常复杂，对过滤以后的目标数据进行特定条件的查询，查询条件可能也非常复杂。在本综合实战案例中，我们使用广播变量，将广播变量分发到各个 Executor 中进行查询过滤，使用 RDD 的 filter 操作，过滤出使用苹果手机搜索网站商品的用户搜索记录。

2) 对用户登录网站的渠道将通过苹果手机登录的记录过滤以后，我们对目标数据构建 Key - Value 类型的 RDD (date#Item#userID, 1L)，Key 值为日期、商品及用户 ID 使用#连接的拼接字符串，Value 值计数为 1，使用 reduceByKey 操作，统计 Key 值的汇总计数，即每用户在每天点击搜索每商品的总次数。

3) 对 reduceByKey 以后的数据记录拆分重新组拼，拆分获取字段 Date、UserID、Item，然后组拼加上汇总统计的搜索次数 count，组成包含 Date、UserID、Item、count 的 Json 字符串，构造 DataFrame。

4) 在 Spark SQL 注册临时表，使用窗口函数 row_number 统计出每用户搜索每商品的前 3 名。

5) Spark SQL 网站搜索结果持久化：可以通过 RDD 直接操作 Mysql，把结果直接放入生产系统数据库 DB 中，通过 Java EE、Web 页面进行可视化展示，提供市场营销人员、仓储调度系统、快递系统、管理决策人员使用。也可以放在 Hive 中，通过 Java EE 使用 JDBC 连接访问 Hive；也可以就放在 Spark SQL 中，通过 Thrift Server 提供 Java EE 使用；这里我们以 JSON 格式保存到磁盘文件系统中。

本案例的具体实现如下：

1) 查询每天每用户点击某商品的次数。

- 初始化 JavaSparkContext 及 HiveContext。通过 sc.textFile 加载用户搜索数据日志文件。
- 使用广播变量 deviceBroadcast，将要过滤的用户终端类型（安卓、苹果、平板）进行广播，使用 filter 方法进行过滤。
- 查询每天每用户点击某商品的次数。对每行的数据按"\t"进行分割，将日期、商品、用户 ID 组成 Key 值，每次计数为 1 次，生成 key - Value 键值对（date#Item#userID, 1）。即每天每用户点击某商品的次数计数为 1。

代码如下：

```
SparkConf conf = new SparkConf().setMaster("local").setAppName("Spark SQLUserlogsHottest");
JavaSparkContext sc = new JavaSparkContext(conf);
SQLContext sqlContext = new HiveContext(sc);
JavaRDD<String> lines = sc.textFile("G:\\Spark SQLData\\Spark SQLUserlogsHot.log");
String device = "iphone";
final Broadcast<String> deviceBroadcast = sc.broadcast(device);//定义广播变量
//使用 filter 方法进行过滤
JavaRDD<String> lineFilter = lines.filter(new Function<String,Boolean>() {
    @Override
    public Boolean call(String s) throws Exception {
        return s.contains(deviceBroadcast.value());//过滤出登录渠道包含苹果手机的记录
    }
```

第8章 Spark SQL综合应用案例

```
});
JavaPairRDD<String,Integer> pairs = lineFilter.mapToPair(new PairFunction<String,String,Integer>
(){
        private static final long serialVersionUID = 1L;
        @Override
        public Tuple2<String,Integer> call(String line) throws Exception {
            String[] splitedLine = line.split("\t");//按"\t"进行分割
            int one = 1;//计数值为1
            String dataanditemanduserid = splitedLine[0] + "#" + splitedLine[2] + "#"
                    + String.valueOf(splitedLine[1]);//组拼成(date#Item#userID)字符串
            return new Tuple2<String,Integer>(String.valueOf(dataanditemanduserid),Inte-
ger.valueOf(one));

        }
});
```

2) 统计每用户在每天点击搜索每商品的总次数。
- 使用 reduceByKey 方法将用户每天点击商品的次数汇总累加。统计出每天用户搜索每商品的搜索次数累计值。
- 将统计结果拼接成 JSON 格式。

代码如下：

```
JavaPairRDD<String,Integer> pairsCount = pairs.reduceByKey(new Function2<Integer,Integer,Inte-
ger>(){
        @Override
        public Integer call(Integer v1,Integer v2) throws Exception {
            return v1 + v2;//相同 key 值的记录,value 进行累加
        }
});

List<Tuple2<String,Integer>> pairsCountRows = pairsCount.collect();
// 动态组拼接为 JSON 格式
List<String> userLogsInformations = new ArrayList<String>();
for(Tuple2<String,Integer> row :pairsCountRows){
        // 按"#"进行分割,拆分三个字段
        String[] rowSplitedLine = row._1.split("#");
        String rowuserID = rowSplitedLine[2];
        String rowitemID = rowSplitedLine[1];
        String rowdateID = rowSplitedLine[0];
        // 拼接 Json 元数据:Date、UserID、Item、Count
        String jsonZip = "{\"Date\":\"" + rowdateID + "\",\"UserID\":\"" + rowuserID + "
\",\"Item\":\"" + rowitemID + "\",\"count\":" + row._2 + "}";
```

```
            userLogsInformations.add(jsonZip);
    }
}
```

3) 在 Spark SQL 中注册临时表,使用窗口函数 row_number 统计出用户搜索每商品的前 3 名。
- 调用 sc.parallelize 方法创建 userLogsInformationsRDD,使用 sqlContext.read().json 加载 Json 数据,构建 DataFrame userLogsInformationsDF。
- 将 DataFrame 注册成为临时表 userlogsInformations。
- 使用 SQL 窗口函数:以子查询的方式完成目标数据的提取,在目标数据内使用窗口函数 row_number 进行分组排序。

(1) 先进行子查询 sub_userlogsInformations,根据用户 ID 分组,按照搜索次数降序排序,查询出用户 ID,商品 ID,搜索次数,排名;

(2) 然后从子查询 sub_userlogsInformations 查询,查询排名前 3 名的用户 ID,商品 ID,搜索次数。

代码如下:

```
//通过内容为 JSON 的 RDD 来构造 DataFrame
    JavaRDD<String> userLogsInformationsRDD = sc.parallelize(userLogsInformations);
    DataFrame userLogsInformationsDF = sqlContext.read().json(userLogsInformationsRDD);
    userLogsInformationsDF.show();
    // 注册成为临时表
    userLogsInformationsDF.registerTempTable("userlogsInformations");
    /*使用子查询的方式完成目标数据的提取,在目标数据内使用窗口函数 row_number 来进行分组排序: PARTITION BY:指定窗口函数分组的 Key; ORDER BY 分组以后进行排序;
    */
    String sqlText = "SELECT UserID,Item,count " + "FROM (" + "SELECT " + "UserID,Item,count," + "row_number() OVER (PARTITION BY UserID ORDER BY count DESC) rank" + " FROM userlogsInformations "
                    + ") sub_userlogsInformations " + "WHERE rank <=3 ";
    System.out.println(sqlText);
    DataFrame userLogsHotResultDF = sqlContext.sql(sqlText);
    userLogsHotResultDF.show();//打印输出 userLogsHotResultDF
    userLogsHotResultDF.write().format("json").save("G://Spark SQLData//Result.json");
//保存到磁盘文件
    while (true) {
    }
```

sqlContext.sql(sqlText)使用开窗函数查询每用户搜索每商品的前 3 名的数据记录,打印输出 userLogsHotResultDF。userLogsHotResultDF.show () 运行结果如下:

```
+---------------+---+----+
```

第8章 Spark SQL综合应用案例

```
|           UserID|    Item|count|
+-----------------+--------+-----+
|a76ff021-049c-4a1...|行车记录仪|  25|
|a76ff021-049c-4a1...|    显卡|  25|
|a76ff021-049c-4a1...|  洗衣液|  23|
|692b1947-8b2e-45e...|   显示器|  31|
|692b1947-8b2e-45e...|   洗衣机|  30|
|692b1947-8b2e-45e...|    小米|  29|
|8768f089-f736-434...|行车记录仪|  33|
|8768f089-f736-434...|   显示器|  29|
|8768f089-f736-434...|  洗衣液|  26|
```

4）将结果以 JSON 格式保存到磁盘文件。

将每用户搜索每商品的前3名的数据保存到本地磁盘文件 G:/Spark SQLData/Result.json，查看 window 系统中的 G:/Spark SQLData/Result.json 目录，目录里面已经生成一批小文件。文件格式包括_SUCCESS、._SUCCESS.crc、.part-r-00000-c4e7c8c9-d08a-443d-9238-e1a13ba1cee1.crc（类似多个文件）、part-r-00018-70d42435-fd8c-409c-882f-3c4f0727d392（类似多个文件）。查看其中一个用户搜索商品的前3名的数据结果文件，如 part-r-00186-c4e7c8c9-d08a-443d-9238-e1a13ba1cee1 的文件内容，显示用户 7371b4bd-8535-461f-a5e2-c4814b2151e1 的搜索前三名的商品为"显卡"，计数35次；"洗衣机"，计数32次；"休闲鞋"计数29次，查询结果如下。

```
{"UserID":"7371b4bd-8535-461f-a5e2-c4814b2151e1","Item":"显卡","count":35}
{"UserID":"7371b4bd-8535-461f-a5e2-c4814b2151e1","Item":"洗衣机","count":32}
{"UserID":"7371b4bd-8535-461f-a5e2-c4814b2151e1","Item":"休闲鞋","count":29}
{"UserID":"a2bccbdf-f0e9-489c-8513-011644cb5cf7","Item":"显卡","count":28}
{"UserID":"a2bccbdf-f0e9-489c-8513-011644cb5cf7","Item":"洗衣液","count":25}
{"UserID":"a2bccbdf-f0e9-489c-8513-011644cb5cf7","Item":"显示器","count":22}
```

5）在 Spark Web UI 页面中查看程序运行情况。

SparkSQLUserlogsHot.Java 使用 Spark 以本地 Local 模式运行，在 Spark SQLUserlogsHot 类的末尾加上 while(true){}循环语句，这样 Spark SQLUserlogsHot.Java 程序一直运行，就可以登录 Spark web 页面 http://127.0.0.1:4040 查看 Spark SQLUserlogsHot 应用运行情况。

登录 Spark Web UI 页面 http://127.0.0.1:4040/jobs/，查询页面显示如图8-3所示。

登录 Spark Web UI 页面 http://127.0.0.1:4040/SQL/，查询 SQL 页面。点击 Detail 列表"==Parsed Logical Plan=="项中右侧的[+details]键，可查看 Parsed Logical Plan 的详细内容，里面包括 Spark SQL 的 Parsed Logical Plan、Analyzed Logical Plan、Optimized Logical Plan、Physical Plan 的执行情况。如图8-4所示。

Job Id	Description	Submitted	Duration	Stages: Succeeded/Total	Tasks (for all stages): Succeeded/Total
7	save at SparkSQLUserlogsHot.java:135	2016/08/17 08:38:53	6 s	2/2	201/201
6	show at SparkSQLUserlogsHot.java:134	2016/08/17 08:38:51	2 s	1/1 (1 skipped)	199/199 (1 skipped)
5	show at SparkSQLUserlogsHot.java:134	2016/08/17 08:38:50	1 s	2/2	2/2
4	show at SparkSQLUserlogsHot.java:117	2016/08/17 08:38:48	0.3 s	1/1	1/1
3	json at SparkSQLUserlogsHot.java:115	2016/08/17 08:38:48	0.1 s	1/1	1/1
2	collect at SparkSQLUserlogsHot.java:88	2016/08/17 08:38:48	0.2 s	2/2	2/2
1	collect at SparkSQLUserlogsHot.java:72	2016/08/17 08:38:47	47 ms	1/1	1/1
0	collect at SparkSQLUserlogsHot.java:51	2016/08/17 08:38:47	0.2 s	1/1	1/1

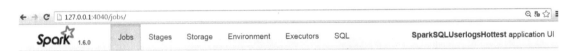

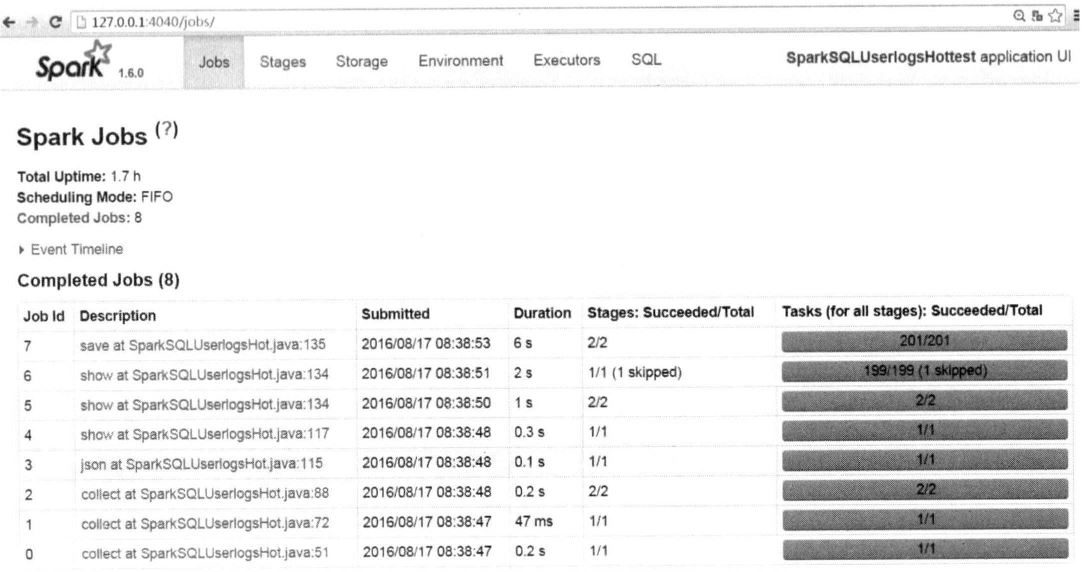

图 8-3　Spark Web UI 页面

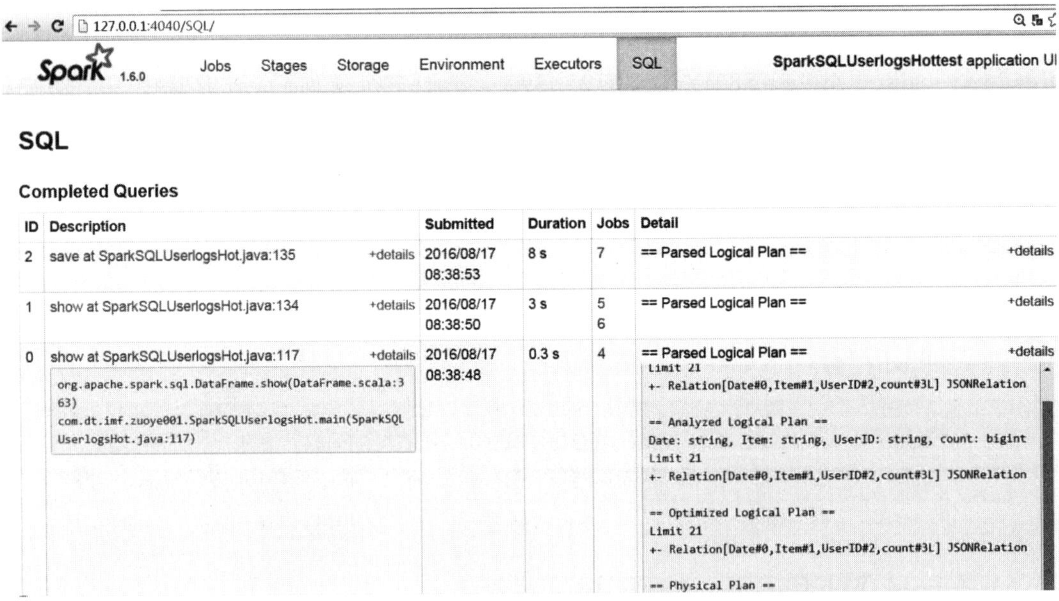

图 8-4　Spark SQL 页面

Section 8.3 本章小结

本章完整的讲述了两个综合企业级应用案例：电子商务网站日志多维度数据分析，电商网站搜索综合排名，归纳并应用了全部 Spark SQL 知识点，是学习 Spark SQL 应用的经典案例，可以使读者对 Spark SQL 有深入的理解。